Elementary Statistics for Economics and Business

Elementary Statistics for Economics and Business

LAWRENCE J. KAPLAN

The City University of New York

Library of Congress Catalog Card Number: 66-10186
Published in cooperation with Landsberger Publishing Corporation
DESIGN BY ANITA DUNCAN

Manufactured in the United States of America

1.987654321

Elementary Statistics for Economics and Business *is one of a series of textbooks published in cooperation with E. K. Georg Landsberger.*

To My Wife

Preface

Professional statisticians and scholars have made significant efforts in bringing technical information to the public. This is evidenced by the large number of statistical texts which come into the market every year. Many of these volumes, however, are beyond the grasp of the average student who takes the basic course.

This book is specifically designed to help this student by starting him cautiously at the shoreline, as any seeker of wisdom, and proceeding slowly toward the depths. The goals have been modest: First, to provide a short, nontechnical introduction to basic statistics; second, to illustrate theory and concepts with practical problems and examples; third, to reduce the volume of note-taking in the classroom by providing a concise, organized body of material; and fourth, to stimulate those with statistical aptitude to more detailed study.

By presenting basic ideas simply and concisely, this book provides the beginner with a set of building blocks which serve as a foundation for understanding more advanced concepts. It has grown out of fifteen years of teaching experience in the college classroom. It is a product of much reading, research, and reflection on statistical concepts and their most effective presentation. Communicating these concepts comprehensively has always presented a challenge. This book, resulting from such challenge, attempts to provide an elementary introduction with clarity and brevity.

The material requires average ability in mathematics with a knowledge of basic arithmetic and algebra. The appendixes provide a review of specific mathematical concepts required for basic statistics. The exercises selected were chosen because it has been my experience that these substantive problems give students a solid grasp of the subject matter.

The book may be utilized as an introductory text by itself or as a supplement to a more detailed text. It represents an honest effort to provide an introduction which will meet the needs of the beginning student. If it should provide a welcome key in opening statistical compounds, it will have accomplished its purpose and satisfied its author.

The author is indebted to the Literary Executor of the late Sir Ronald A. Fisher, F.R.S., Cambridge, to Dr. Frank Yates, F.R.S., Rothamsted, and to Messrs. Oliver & Boyd Ltd., Edinburgh, for permission to reprint Tables V and VI of the Appendix from their book *Statistical Tables for Biological, Agricultural and Medical Research.*

The author is also grateful to Professor Earl S. Paul, Jr. of Rensselaer Polytechnic Institute and Dr. Helen Hunter of Swarthmore College for reviewing the manuscript and submitting their valuable comments and suggestions. He also wishes to thank Dr. Jerome Ozer and Miss Susan Sperling of the Pitman Publishing Corporation for their assistance in the preparation of this manuscript.

LAWRENCE J. KAPLAN

New York, New York
July 1966

Contents

Elementary Statistics for Economics and Business

1

Introduction

1.1 THE NATURE OF STATISTICS

The study of statistics is actually the study of statistical theory and methods. *Statistics*, therefore, may be defined as the principles and methods used in the collection, presentation, analysis, and interpretation of numerical data.

While most professional statisticians are concerned primarily with these basic functions—collection, presentation, analysis, and interpretation of data—a growing number of statisticians are applying the principles of probability in the area of decision-making. In this approach, statistics may be defined as the art and the science of collecting, analyzing, and making inferences from data, or perhaps the art and science of decision-making in the face of uncertainty.

In either approach, statistics provides the tools for extracting basic truths which often lie hidden in a mass of data. It sets forth the procedures for collecting data, recording it in a systematic way, and classifying individual items or observations. This permits analysis, interpretation, and the making of inferences.

The study of elementary statistics requires a knowledge of algebra, skill in the use of arithmetic, familiarity with simple equations, and the use of simple mathematical tables.

Elementary statistics excludes the derivation of the mathematical formulas developed in advanced mathematics courses. Applied statistics accepts these formulas and uses them as tools in solving problems.

1.2 REASONS FOR STUDYING STATISTICS

The reasons for studying statistics are, first, to develop statistical literacy, and thus to make each student an intelligent user of statistics. The ability to understand and evaluate numerical data is a basic skill required of the citizens in a democracy. And, a second reason is to stimulate an appreciation of the variety of applications of statistics in contemporary life.

1.3 APPLICATIONS OF STATISTICS

Science: Scientific laws describe how nature has behaved within specific limits, and how it is likely to behave under similar conditions. Statistics determines what these limits are, under a given set of conditions, and develops the probability of the recurrence of a given set of events, based upon frequency and regularity of occurrence in the past.

Government: Financial officials of government apply statistical methods in measuring the gross national product, savings, investments, and changes in the purchasing power of money. Moreover, economists calculate tax rates, and tax income, against governmental financial needs and expenditures.

Business: Business executives project long-run capital requirements, forecast sales and production, and analyze quality of production to improve their product and to achieve the best use of personnel and materials. The stock market uses statistics of its daily operations to guide the trading public.

Politics: Elected officials and those seeking public office rely upon statistical research to make decisions on where to concentrate money and manpower to win votes. They rely on statistical polls to evaluate citizen support, just as polls are used in the entertainment field to estimate popular appeal.

Miscellaneous: Accurate statistics of mortality and morbidity enables actuaries to compute insurance premiums which are reasonable and profitable to the company. Public health statistics play a vital role in protecting the well-being of the nation's population.

1.4 MISUSES OF STATISTICS

The user of statistics must carefully evaluate the data he is using in order to interpret them correctly. To do this, however, the user must be aware of the many pitfalls involved in using statistical data. The principal objective of this section is to indicate some of the more common misuses of statistics. This knowledge should increase one's desire to study statistical methods, which follow, thus insuring a more effective and skillful use of statistics. Several types of misuse are presented below.

Built-in bias: These statistics serve a vested interest. It is easy to detect this type of bias in advertisements which state that 3 out of 4 doctors recommend a particular product or that statistics prove a given product is superior to a competitor's product.

Unjustified generalizations: A common mistake in statistical reasoning is to draw a conclusion on the basis of a sample which is too small, or one which is not representative of the population from which the sample was selected. A well-known example relates to the presidential campaign of 1936 in which the election of Republican candidate Alfred Landon over the Democrat F. D. Roosevelt was predicted by the *Literary Digest* magazine. In that year, Roosevelt carried every state in the union except Maine and Vermont. The forecasting error was due to a biased sample. The entire question of sample selection is considered in Chapter 7.

Noncomparable data: Frequently, a comparison is made between two things which are not alike. Sometimes this is referred to as comparing apples and oranges. For example, the Consumer Price Index, prepared by the U.S. Department of Labor's Bureau of Labor Statistics, cannot be compared with a similar index prepared by a private, nonprofit organization. The number of items included, the specifications, the weighting technique, and sources of data are different.

Percentages versus absolute numbers: Percentages are used widely in statistical analysis. However, they are often improperly calculated by using an incorrect base, or by failing to subtract 100 per cent in computing percentage increases; or the nature of the comparison may not have been correctly understood. A classic example of the last error concerns the $33\frac{1}{3}$ per cent of the women students at Johns Hopkins University who married faculty members. This appeared significant until it was learned that only three women students were enrolled at the time and one had married a faculty member.

Changing variables: Many individuals are interested in forecasting, and extrapolate trend data over long periods on the assumption that underlying variables will remain the same. For example, a population forecast made in Washington, D.C., in 1865 predicted that the population of the United States would be almost 252 million in 1930. Actually, it was 122.8 million. The forecast was based on the assumption that the rate of increase of the United States population would be the same between 1860 and 1930 as it had been between 1790 and 1860.

Disregard of dispersion: A measure of variation or dispersion must accompany an average measure. The story is told of a nonswimmer who was drowned in a river that had an average depth of 3 feet in the dry period. While he could safely wade in 3 feet of water, at some points the water depth was 10 to 15 feet.

Technical errors: These errors are basically errors in calculation. For example, a corporation was controlled by 5 individuals each of whom had 100 votes, or 500 votes among them. The remaining 100 votes were controlled by 50 individuals. The erroneous conclusion reached was that the average number of votes per individual was almost 11, or 600 ÷ 55.

These examples of misuses of statistics are not exhaustive. They merely represent certain basic types of statistical misrepresentations. For a humorous treatment of this subject, see Darrell Huff, *How To Lie With Statistics*, cited in the Bibliography.

EXERCISES

1. Define or explain the following:

   Statistics

   Misuses of statistics

2. Read Appendix D, Selected Sources of Economic and Business Data. In one paragraph for each source, describe the content of the publication.

3. The American Statistical Association is one of the oldest professional organizations in the United States. It was founded November 27, 1839. Its present membership is about 8,500 persons. Refer to a recent issue of the *Journal of the American Statistical Association.* Read one article, and, in one paragraph, summarize its content.

4. Using newspapers, magazines, or any other printed material, select and explain a misuse of statistics.

2

Presentation of Data: Tables and Graphs

Presentation of data is one of the most important considerations in statistical work. The most effective way of presenting statistical or quantitative data is through the use of tables and graphs. The primary function of a table is to organize and classify data. A graph presents these data visually. Visual presentation generally makes quantitative data more easily understood.

The statistical table is the most important form for presenting data. The graph is used primarily to present statistical data so that they may be understood quickly and clearly. The adage that "a picture is worth ten thousand words" is true and explains the popularity of picture magazines and books.

This chapter analyzes the uses of tables, and then presents a definition, the classes of tables, and the parts of a table. It defines a graph, illustrates its parts, classifies the types of graphs, and illustrates each type. It highlights the uses of line graphs and bar graphs, and analyzes in detail the semilogarithmic line diagram. It explains how to interpret the semilogarithmic graph, and illustrates the use of this graph by analyzing some population trends. The chapter is concluded with rules for graph construction and the pitfalls to be avoided in constructing them.

2.1 TABLES

Uses of Tables

A well-constructed table has several important uses which make it valuable for the presentation of data. Among its major uses are the following:

1. It communicates information far better than textual material.
2. It presents facts clearly and concisely, eliminating the need for wordy explanation.
3. It facilitates comparative analyses of data.
4. It simplifies reference to data.
5. It facilitates interpretation of data.

DEFINITION: A *table* is a concise and orderly presentation of related statistical data or other information in columns and rows.

Classes of Tables

Tables are of two general types: reference, or primary, tables, and text, or derived, tables.

The *reference* or *primary table* presents original data in tabular form to facilitate easy reference. For example, the Bureau of the Census of the U.S. Department of Commerce publishes hundreds of reference tables upon completion of one of its periodic censuses. These tables contain many columns and rows of data.

The *text* or *derived table* is designed to highlight a particular set of facts in simple form. It answers a specific question and is easier to grasp than the reference table. It is frequently woven into a textual analysis.

Parts of a Table

Professional statisticians, responsible for tabular presentation, generally require that a table be constructed using the format shown in the table model. Compare it with Table 2.1.

Title: The title must explain clearly and briefly what information is being presented. A complete title must answer the following questions:

What is being presented

How the data are classified

Where or to which place the data refer

When or to which time period or date the data refer

Headnote: The headnote is a further explanation of some part of the title. For example, it may indicate that the money reference in the title refers to thousands of dollars.

TABLE MODEL

TITLE

Headnote, if any

	Caption					
Stubhead	Column head	Column head	Column head	Column head	Column head	Column head
Stub entries			Body			

Footnote(s), if any
Source note

The stubhead is the caption of the first column.

Stub entries: Each entry in the first column is an explanation of a row.

The caption is a general description of all the column heads.

The column head describes the data in each column.

Body: The body of the table contains the actual data being presented. It consists of cells, each of which is described by the column and row. It is similar to the square of a checkerboard.

Footnote: A footnote is used to explain, describe, or qualify a particular row, column, or cell in the table. Footnote reference symbols may be letters, numbers, or symbols.

Source: The source note should indicate author, title, publisher, date, volume and page. It is placed at the bottom of the table, below the footnotes, if any. The source of the data in a table must always be shown.

2.2 INTRODUCTION TO GRAPHS

Graphs are widely used by business and government statisticians to portray research findings visually. Graphs are used in the analysis and interpretation of statistical data as well as to supplement statistical tables in the presentation of data.

DEFINITION: A graph is a device used in statistics to present data visually, thus highlighting basic facts and relationships.

TABLE 2.1 NEW HOUSING UNITS STARTED, TOTAL AND NONFARM, BY OWNERSHIP AND TYPE OF STRUCTURE; ANNUALLY 1959–1961 AND MONTHLY 1961 AND 1962

(thousands of units)

Period	Total (including Farm).				Nonfarm							
	Private and Public				Total, Private and Public				Private			
	Total	*One-family*	*Two-family*	*Three-family or more*	*Total*	*One-family*	*Two-family*	*Three-family or more*	*Total*	*One-family*	*Two-family*	*Three-family or more*
Annual totals:												
1959	1,553.5	1,250.7	58.5	244.3	1,531.3	1,228.7	58.5	244.1	1,494.6	1,211.7	55.8	227.1
1960	1,296.0	1,008.8	50.5	236.8	1,274.0	986.6	50.5	236.8	1,230.1	972.3	43.8	213.6
1961	*1,355.4	*980.1	*50.0	*325.0	*1,327.2	*952.1	*50.0	*325.0	*1,275.5	*937.5	*43.7	*294.4
First 5 months:												
1961	509.2	371.7	21.4	115.9	497.3	359.9	21.4	115.9	475.9	351.5	18.3	106.2
1962	*584.4	*396.2	*22.4	*165.7	*576.8	*388.6	*22.4	*165.7	*563.8	*387.2	*20.3	*156.3
Monthly:												
1961: April	115.3	85.4	4.2	25.6	113.0	83.1	4.2	25.6	108.7	81.4	3.7	23.6
May	130.7	97.9	5.4	27.4	128.3	95.6	5.4	27.4	124.2	94.1	4.4	25.7
June	138.3	100.6	4.3	33.3	135.3	97.6	4.3	33.3	129.5	96.9	3.9	28.6
July	128.5	97.6	4.3	26.6	126.0	95.1	4.3	26.6	122.7	93.7	3.8	25.2
August	130.1	96.1	3.9	30.1	127.3	93.3	3.9	30.1	124.2	92.0	3.3	28.8
September	128.2	91.5	4.3	32.4	126.5	89.8	4.3	32.5	120.7	89.1	3.9	27.9
October	128.9	94.1	5.0	29.8	126.4	91.7	5.0	29.7	121.5	89.9	4.3	27.3
November	105.5	74.1	3.7	27.7	103.8	72.4	3.7	27.7	100.8	72.3	3.3	25.2
December	86.7	54.4	3.1	29.2	84.5	52.3	3.1	29.2	80.2	52.1	2.9	25.2
1962: January	83.0	54.4	3.1	25.5	81.7	53.1	3.1	25.5	79.3	53.1	2.8	23.4
February	77.8	53.8	3.0	21.0	76.7	52.7	3.0	21.0	75.3	52.6	2.5	20.2
March	117.9	79.8	5.1	33.0	116.3	78.1	5.1	33.0	113.8	78.0	4.6	31.2
April	*151.6	*101.7	*5.8	*44.1	*149.5	*99.6	*5.8	*44.1	*144.9	*98.9	*5.4	*40.6
May	*154.1	*106.5	*5.4	*42.1	*152.6	*105.1	*5.4	*42.1	*150.5	*104.6	*5.0	*40.9

Note: Components may not equal totals due to rounding.
* Preliminary.
SOURCE: U.S. Department of Commerce, *Construction Report*, C20-37, July 1962.

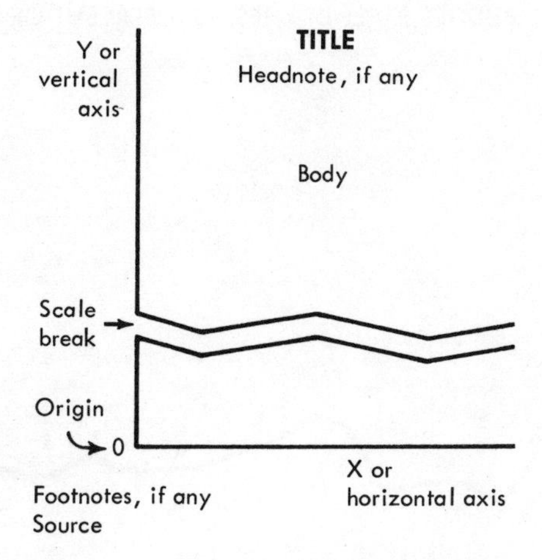

Fig. 2.1 Illustration: Parts of a Graph

Parts of a Graph

Title: The title of a graph must answer the same questions as the title of a table, such as, what? how? where? and when? Generally, a table and graph presenting the same data are placed in close proximity. The title of both table and graph would be the same.

X and Y axes: The horizontal axis or line is known as the *X* axis and the vertical axis or line is known as the *Y* axis.

Origin: The point at which the two axes converge is known as the origin and is assigned the value zero (0). The origin must always be shown.

Scale break: If the scale entries on the vertical axis require so much space that the *Y* axis is unnecessarily elongated, a portion of the scale entries may be omitted. This omission is indicated by a scale break. (It is illustrated in Figure 2.6 below.)

Body: The body of the graph contains the lines or bars representing the data.

Footnote: A footnote is used to explain, describe, or qualify a particular item on the graph.

Source: The source note should indicate author, title, publisher, date, volume and page. If the data of both the table and graph are the same, the source note would be the same. The source must always be shown.

2.3 TYPES OF GRAPHS

1. Line graphs (see Figures 2.2–2.4)
   a. Arithmetic line diagrams (see frequency polygon, Chapter 3)
   b. Semilogarithmic (or ratio) line diagrams

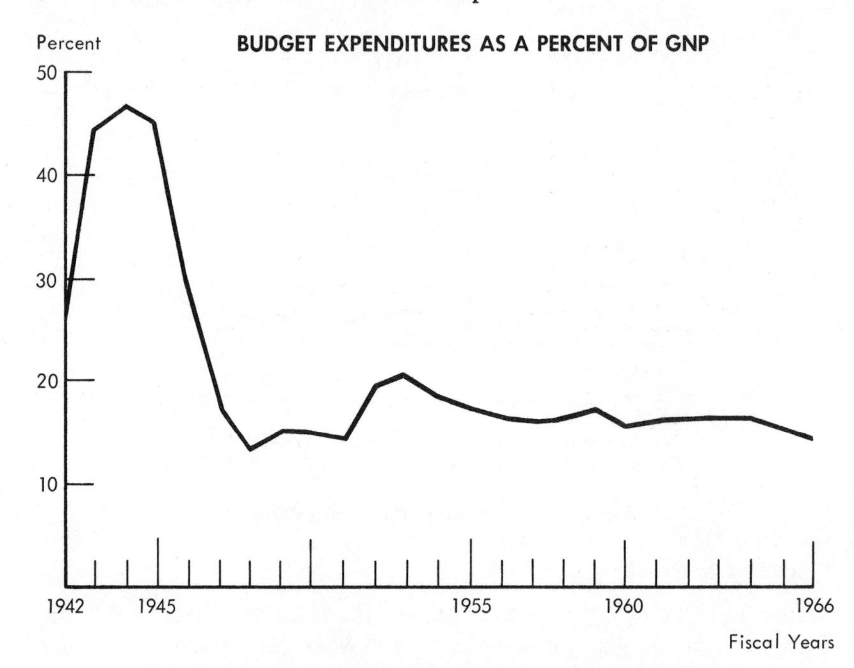

Fig. 2.2 Arithmetic Line Diagram: Single-line Graph
SOURCE: Executive Office of the President, Bureau of the Budget.

2. Geometric forms (see Figures 2.5–2.10)
   a. Bar charts (see histogram, Chapter 3)
   b. Area diagrams
   c. Volume diagrams: cubes, spheres, cylinders
3. Statistical maps (Figure 2.11)
4. Pictographs (Figure 2.12)

Examples of each of these types of graphs follow. Because it is more complex than the other graphs, the semilogarithmic line diagram is discussed last, in Section 2.4 (see page 17). Note that a bar chart or bar graph is also known as a histogram. The bars may be vertical or horizontal. The types are illustrated in Figures 2.6, 2.7, and 2.8.

In Figure 2.7, each bar is equal to 100 per cent. This type of graph shows the relative change in importance of the various component parts, which is very important in statistical analysis. The percentages of each component for each period are plotted on the graph.

Comparison Between Line Graphs and Bar Graphs

1. While the arithmetic line diagram is the most important type for the presentation of data, bar graphs are preferable to line graphs in portraying a relatively few values of one or two series.

VALUE OF NEW CONSTRUCTION PUT IN PLACE, SEASONALLY ADJUSTED ANNUAL RATE

Billions of dollars

60 50 40 30 20 10 0

New Series [a]

TOTAL NEW CONSTRUCTION

TOTAL PRIVATE CONSTRUCTION

PRIVATE RESIDENTIAL CONSTRUCTION

TOTAL PUBLIC CONSTRUCTION

1954 1955 1956 1957 1958 1959 1960 1961 1962 1963

Fig. 2.3 Arithmetic Line Diagram: Multi-line Graph

SOURCE: U.S. Department of Commerce, *Construction Report*, C30-47, June 1963.

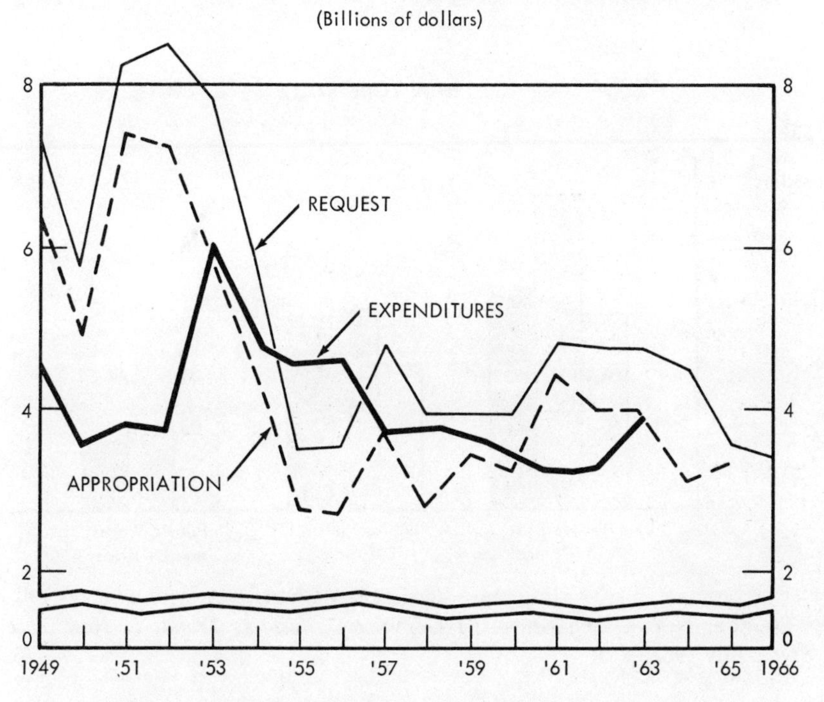

Fig. 2.4 Multi-line Graph Illustrating Use of Scale Break

SOURCE: *New York Times*, February 7, 1965.

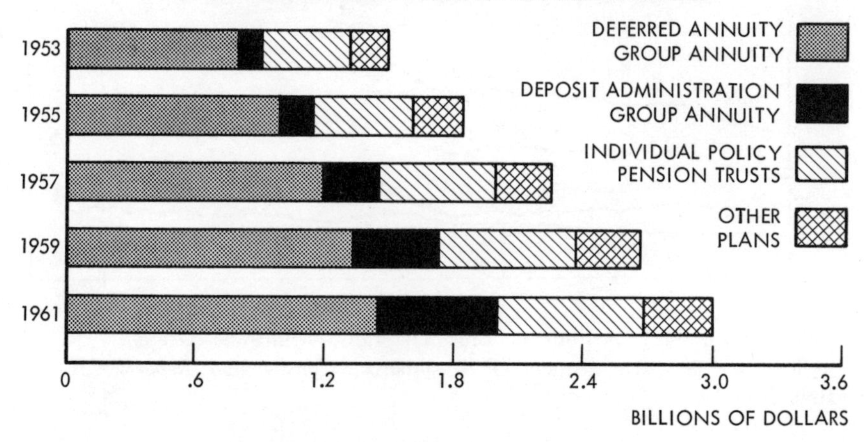

Fig. 2.5 Horizontal Bar Chart
SOURCE: Institute of Life Insurance, *Life Insurance Fact Book*, 1962, p. 36.

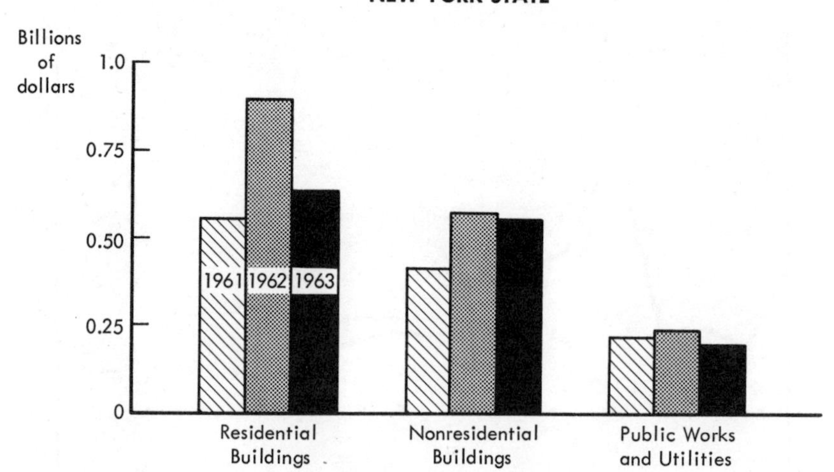

Fig. 2.6 Vertical Bar Chart
SOURCE: N.Y.S. Department of Commerce, *Business Trends in New York State*, July 1963, p. 3.

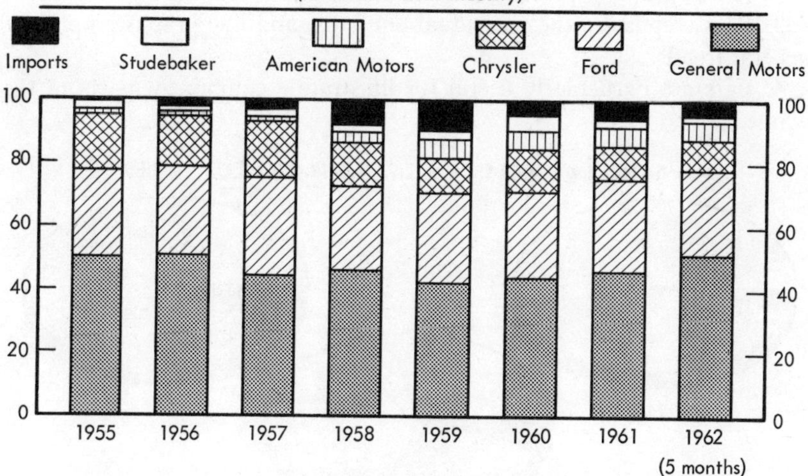

Fig. 2.7 100 Per Cent Component Vertical Bar Graph
SOURCE: *New York Times*, July 29, 1962.

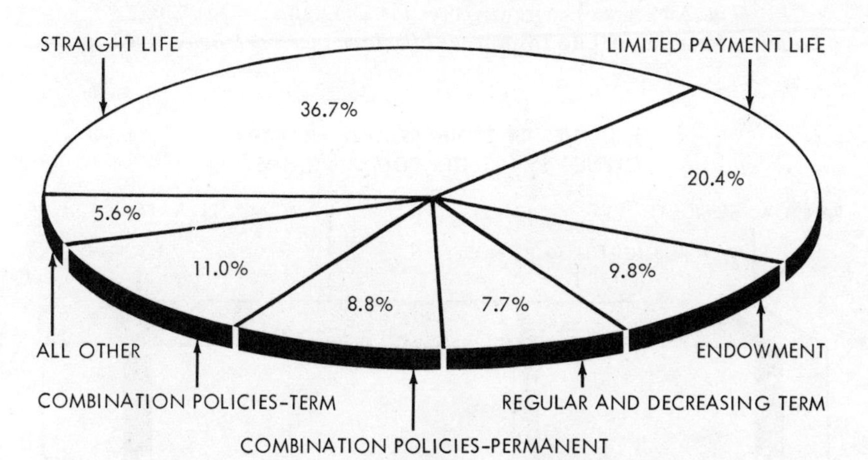

Fig. 2.8 Area Diagram: Pie Chart
SOURCE: Institute of Life Insurance, *Life Insurance Fact Book*, 1962, p. 15.

2. A line graph is preferable when many series are to be presented.

3. Bars emphasize the individual amounts, and line graphs emphasize general trends.

4. Bars are particularly useful for illustrating component parts of the whole.

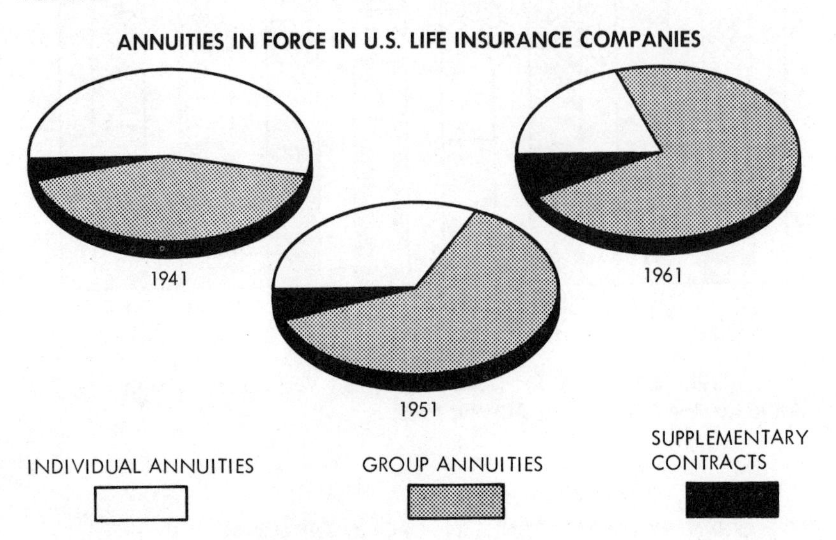

Fig. 2.9 Area Diagram: Three Pie Diagrams in One Chart
SOURCE: Institute of Life Insurance, *Life Insurance Fact Book*, 1962, p. 34.

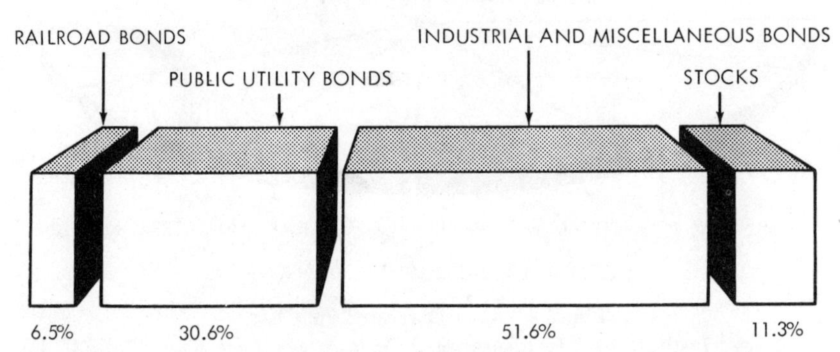

Fig. 2.10 Volume Diagram
SOURCE: Institute of Life Insurance, *Life Insurance Fact Book*, 1962, p. 77.

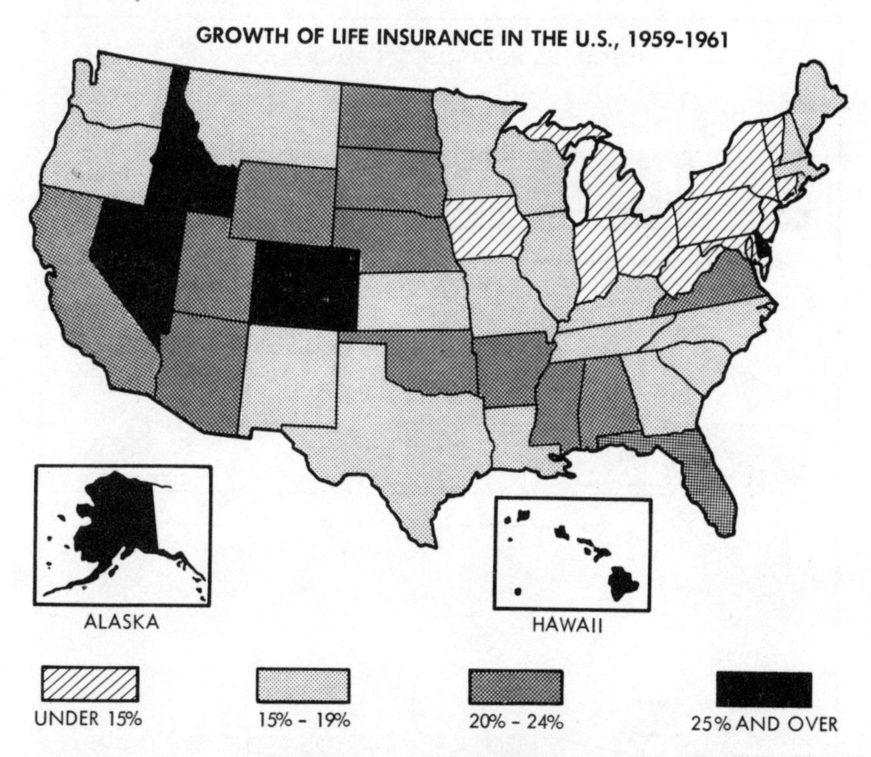

Fig. 2.11 Statistical Map

SOURCE: Institute of Life Insurance, *Life Insurance Fact Book*, 1962, p. 13.

5. Bars are usually set up vertically for both time and size distributions. For qualitative and geographic comparisons the bars are set horizontally.

2.4 SEMILOGARITHMIC (OR RATIO) LINE DIAGRAMS

The second type of line graph is the semilogarithmic, or ratio, chart. This type of graph has two major applications. First, it is used to highlight percentage changes in data, or the rate of change, between one period and another. Second, it is used to compare rates of change in two series of data in different units of measure, such as, inches versus pounds, or two series in which one is counted in thousands and the other in millions.

The difference between the arithmetic line diagram and the semilogarithmic line diagram may be illustrated by comparing an arithmetic progression and a geometric progression.

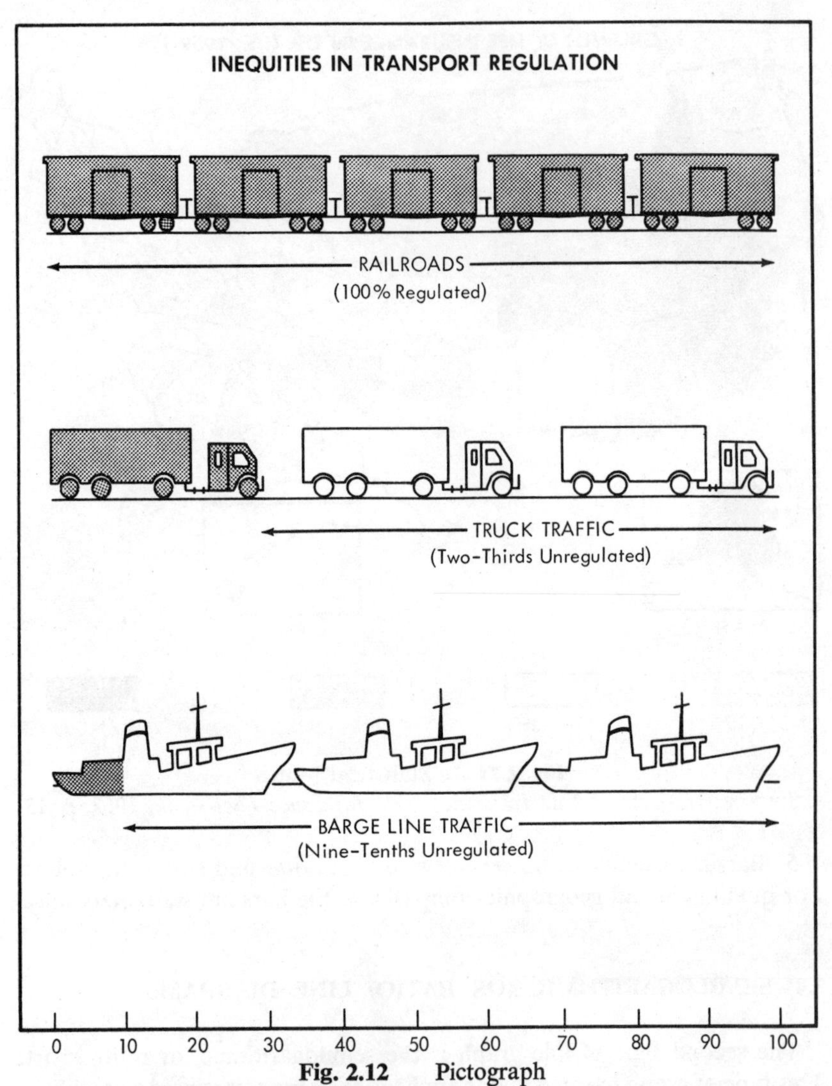

Fig. 2.12 Pictograph

SOURCE: Association of American Railroads, *The Gathering Transportation Storm*, Washington, D.C., 1962, p. 16.

ARITHMETIC PROGRESSION:

Definition: An *arithmetic progression* is a series of numbers which increases or decreases by a constant amount.

Example: 2,4,6,8,10,12,14,16,18,20. The increase from one number to the next is *plus* 2.

The graph of an arithmetic progression: On arithmetic graph paper, an arithmetic progression yields a straight line.

GEOMETRIC PROGRESSION:

Definition: A *geometric progression* is a series of numbers which increases or decreases by a constant percentage.

Example: 2,4,8,16,32,64,128,256,512. The percentage increase from one number to the next is 100 per cent.

The graph of a geometric progression: The geometric progression yields a straight line on semilogarithmic graph paper and a curved line on arithmetic graph paper.

Characteristics of Semilogarithmic Graph Paper

1. The *X* scale is arithmetic, and the *Y* scale is logarithmic.
2. A geometric progression on semilogarithmic paper will yield a straight line.
3. Equal distances on the logarithmic scale represent equal *percentage* changes while on arithmetic graph paper equal distances represent equal *amounts* of change.
4. Semilogarithmic graph paper consists of one or more sections. Each section is called a *cycle* or a *phase*.

Labeling the Vertical or Logarithmic Scale (see Figure 2.13)

1. The values on the vertical scale begin at the bottom of the first cycle, where any positive number may be assigned.
2. The value placed at the top of the first cycle will be ten times the value at the bottom of that cycle. Therefore, the logarithmic scale cannot begin with a zero value.
3. The value placed at the top of the second cycle will be ten times the value at the bottom of that cycle.

Understanding the Semilogarithmic Graph

1. The slope of the line on a semilogarithmic graph indicates the percentage change between any two points in time. The steeper the slope the greater is the percentage change.
2. A straight line indicates a constant percentage [of] change from one period to the next. A curved line indicates a varying rate of percentage change between periods.
3. Equal vertical distances on the semilogarithmic graph indicate equal percentage changes. For example, the distance between 8 and 16, if these numbers appear on the vertical scale, and between 16 and 32 would be equal. In each case, the rate of change is 100 per cent.

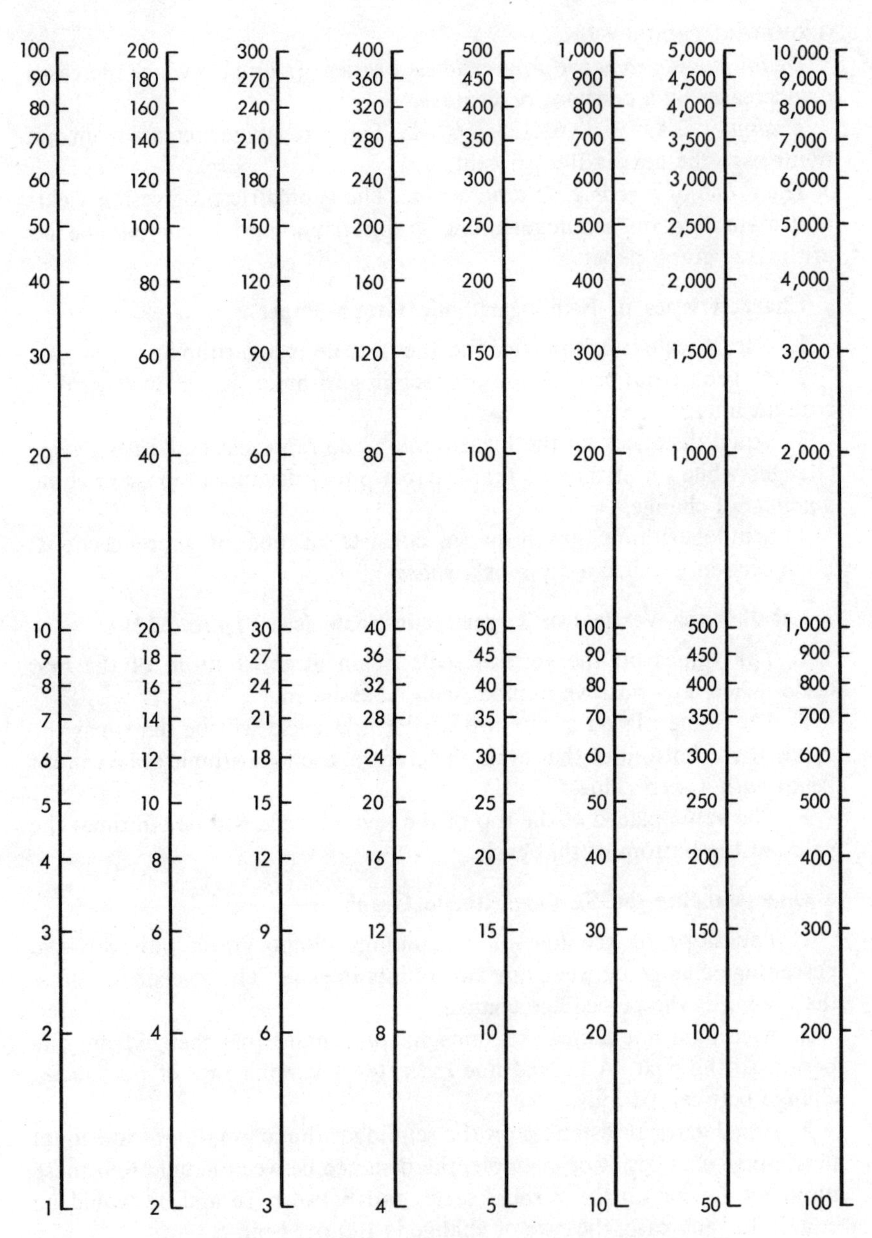

Fig. 2.13 Examples of Semilogarithmic Scale Values

Interpreting the Semilogarithmic Graph

By observing the movement of a curve on a ratio chart, the statistician can determine whether or not a series is maintaining its rate of gain. Seven possibilities may be observed for a data series:

1. Increasing at a decreasing percentage rate
2. Increasing at a constant percentage rate
3. Increasing at an increasing percentage rate
4. No change
5. Decreasing at an increasing percentage rate
6. Decreasing at a constant percentage rate
7. Decreasing at decreasing percentage rate

These possibilities are illustrated in Figure 2.14.

EXAMPLE: Plot the data in Table 2.2 on a sheet of semilogarithmic graph paper. The result is shown in Figure 2.15.

TABLE 2.2 POPULATION OF THE UNITED STATES, NEW YORK CITY, AND QUEENS COUNTY, N.Y., CENSUS YEARS, 1890–1960

	Population		
Year	United States	New York City	Queens County, N.Y.
1890	62,947,714	2,507,414	87,050
1900	75,994,575	3,437,202	152,999
1910	91,972,266	4,766,883	284,041
1920	105,710,620	5,620,048	489,042
1930	122,775,046	6.930,446	1,079,129
1940	131,669,275	7,454,995	1,297,634
1950	150,697,361	7,891,957	1,550,849
1960	178,464,236	7,781,984	1,809,578

SOURCE: U.S. Department of Commerce, Bureau of the Census, *Census of Population*, 1890–1960.

Limitations of Semilogarithmic Graphs

Semilogarithmic graphs have some limitations in data presentation which restrict their application. These include the following:

1. Although absolute magnitudes may be read from the Y axis, they fail to provide a visual picture of these magnitudes as a distance above the base line.

2. Semilogarithmic graphs are difficult for laymen to understand, and therefore should be used only when a simpler type of graph is inadequate.

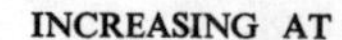

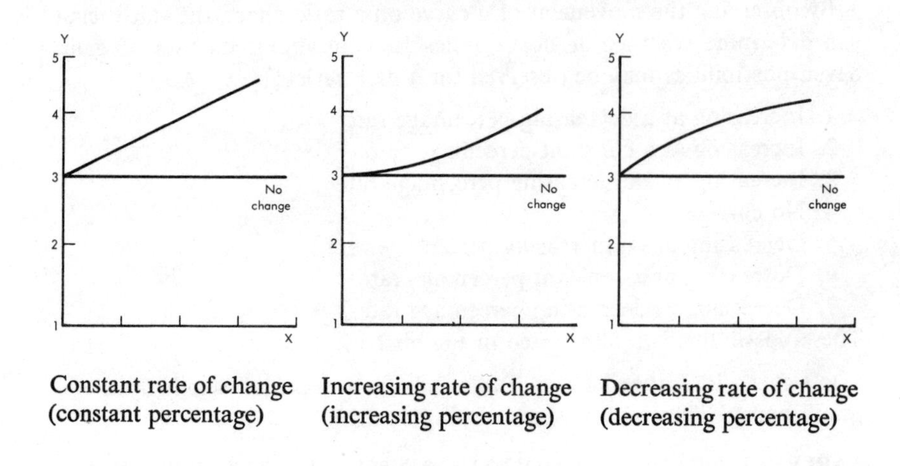

DECREASING AT

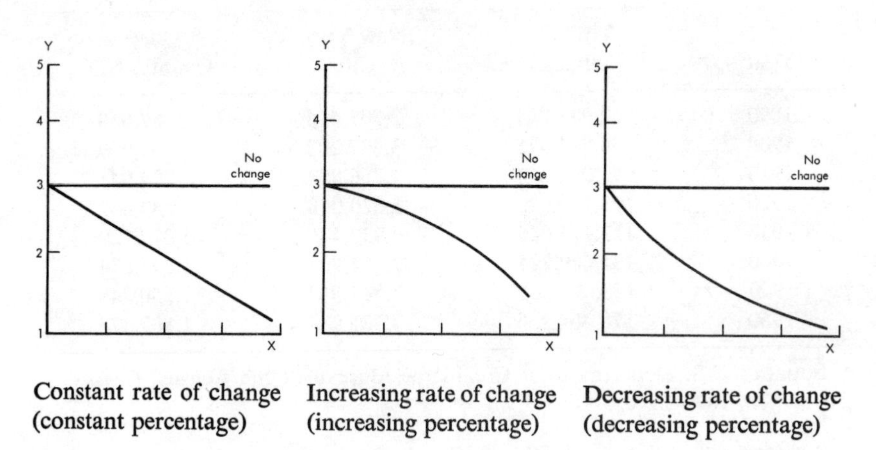

Fig. 2.14 Interpreting the Semilogarithmic Graph

3. Because of the nature of the graph, the *Y* axis cannot show zero or negative values.

4. Sometimes a too wide range of absolute values is condensed into the *Y* axis of a semilogarithmic graph. If only relative movements are of

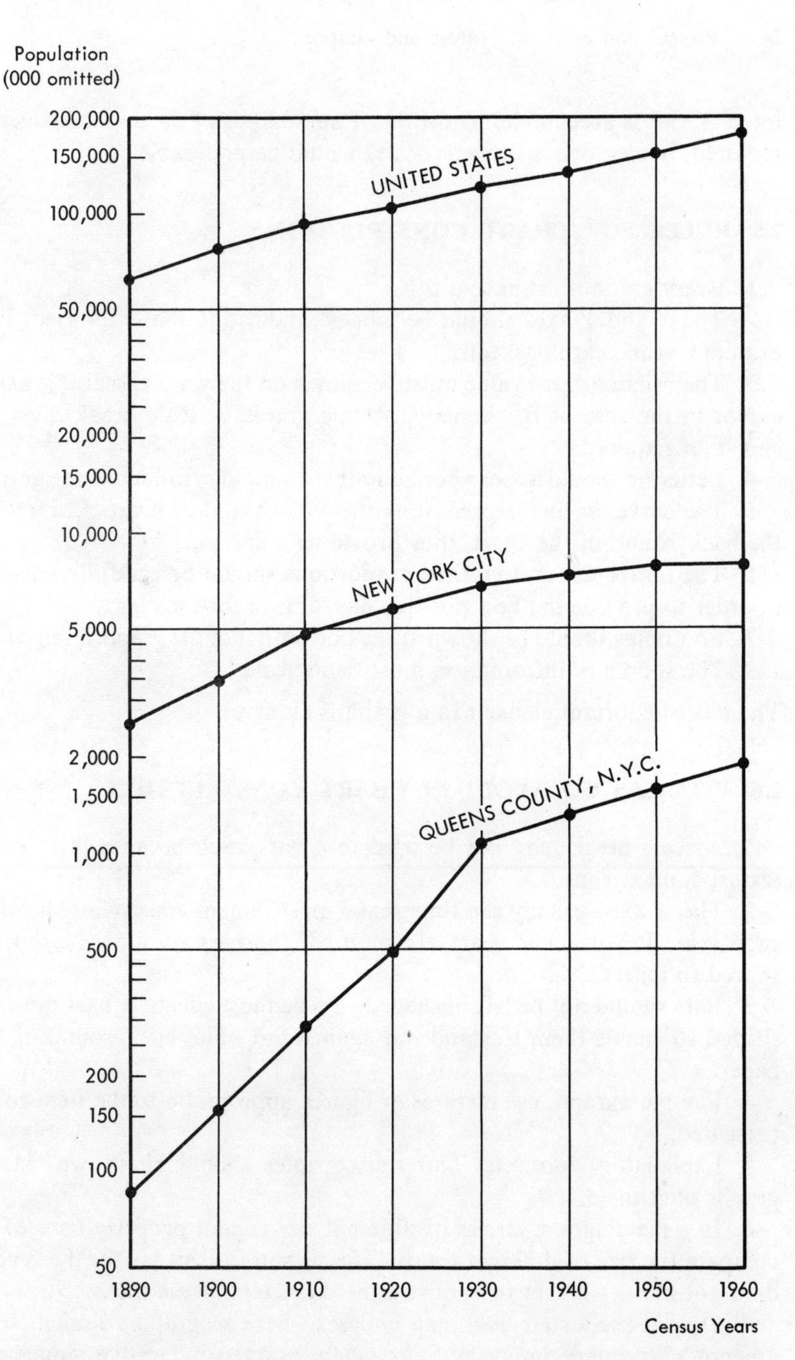

Fig. 2.15 Population of the United States, New York City, and Queens County, N.Y., Census Years, 1890–1960

interest, this is acceptable. However, if an analysis of absolute changes is required, one or two arithmetic scales would be preferable.

2.5 RULES FOR CHART CONSTRUCTION

1. Every graph must have a title.
2. The *X* and *Y* axes should be labeled, indicating the units used. For example, years, earnings, tons.
3. The origin or zero value must be shown on the vertical scale (*Y* axis), except in the case of the semilogarithmic graph. A scale break must be shown, if required.
4. Lettering should appear horizontally to simplify reading of the graph.
5. The curve, or line, representing the data should contrast clearly with the background of the chart, thus providing emphasis.
6. The horizontal and vertical proportions should be carefully selected in order to provide the best possible appearance for the chart.
7. Footnotes should be shown at the bottom left of the graph, if required.
8. The source of information must be indicated.

The most important element in a graph is clarity.

2.6 PITFALLS TO AVOID IN CHART CONSTRUCTION

1. A scale break may not be used in a bar graph because this would distort comparisons.
2. The *X* axis, usually the time scale, must remain consistent, showing each year. If particular years are omitted, the bars should be properly spaced to reflect this.
3. Bars should not be left unshaded. To be most effective, bars must be shaded to enable them to stand out against the white background of the paper.
4. In a pictograph, use pictures or figures appropriate to the item to be presented.
5. Explanatory footnotes and source notes should be shown. Many graphs omit these.
6. In a pie diagram, circles of different size cannot properly be used to compare the size of different totals. The reason for this is that the reader does not know whether to compare the diameters or the areas.
7. Do not use a statistical map in cases where geographic comparisons are not of primary importance or where exact comparative values are required. Bar graphs are preferable for accurate portrayal of size relationships.

8. Do not use a statistical dot map if each dot represents too large an amount. The dots would be so scattered that no apparent density would show up on the map. Conversely, if the unit selected is too small, certain areas of the map will show up completely black. It is not necessary to be able to count the dots, but the unit should be carefully selected so that the effect of density may be clearly visible.

9. Do not use a statistical dot map to show comparisons of ratios, rates, and percentages. Use instead gradations of shading to represent the stratum of each geographic area. These gradations of shading include, for example, light stipling, hatching, crosshatching, and solid black. The lightest shading should indicate the smallest ratio. Degrees of density should move from lightest shading to solid black.

10. One of the most serious pitfalls in graphic presentation is the desire to portray too much on a single graph. This negates the primary objective, namely, clarity.

EXERCISES

1. Define or explain the following:

   Area graph
   Arithmetic line graph
   Arithmetic progression
   Bar graph
   Body
   Caption
   Column head
   Footnote
   Geometric progression
   Graph
   Headnote
   Horizontal axis
   Multi-line graph
   Origin
   Pictograph
   Reference or primary table
   Scale break
   Semilogarithmic graph
   Source
   Statistical map
   Stub entry
   Stubhead
   Table
   Text or derived table
   Title
   Vertical axis
   Volume graph
   X axis
   Y axis

2. By applying the principles of tabular presentation, construct a table of the following data prepared by the U.S. Securities and Exchange Commission and the New York Stock Exchange on the number of shares of stock sold on the New York Stock Exchange, monthly averages, in millions of shares: 1956, 65.3; 1957, 66.5; 1958, 76.8; 1959, 86.6; 1960, 79.9; 1961, 107.7; 1962, 99.0; 1963, 112.5; 1964, 124.0.

3. Using the data given in Exercise 2 above, construct an arithmetic line diagram.

TABLE 2.3 LUNG CANCER DEATH RATES FOR MEN PER 100,000 POPULATION, UNITED STATES, 1950–62

Year	Male death rate
1950	18.4
1951	19.6
1952	21.0
1953	23.1
1954	24.0
1955	25.6
1956	27.6
1957	28.8
1958	29.8
1959	31.2
1960	31.8
1961	33.2
1962	35.0

SOURCE: U.S. National Vital Statistics Division, National Center for Health Statistics; and the American Cancer Society, New York, New York.

TABLE 2.4 MARKET VALUE OF LISTED STOCKS AND TOTAL SHARES SOLD ON THE NEW YORK STOCK EXCHANGE, MONTHLY AVERAGES, 1956–64 (Billions of dollars)

	Market Value	
Year	Listed stocks	Total shares sold
1956	$216.1	$2.5
1957	213.1	2.3
1958	232.1	2.7
1959	295.4	3.6
1960	291.5	3.2
1961	358.9	4.4
1962	339.3	3.9
1963	386.6	4.6
1964	454.1	5.0

SOURCE: U.S. Securities and Exchange Commission and the New York Stock Exchange.

4. Construct a vertical bar chart of the data in Table 2.3.

5. Using the data in Table 2.4 construct an arithmetic line diagram of the market value of listed stocks on the New York Stock Exchange and a horizontal bar chart of total shares sold.

6. Construct a component part bar chart using the data in Table 2.5.

TABLE 2.5 PERSONAL CONSUMPTION EXPENDITURES IN THE UNITED STATES, BY MAJOR TYPE, 1959–64

(Billions of dollars)

Major type	Year					
	1959	1960	1961	1962	1963	1964
Durable goods[a]	$ 43.6	$ 44.8	$ 43.7	$ 48.4	$ 52.1	$ 57.0
Nondurable goods[b]	147.1	151.8	155.2	162.0	167.5	177.3
Services[c]	122.8	131.9	139.1	146.4	155.3	165.1
Total[d]	$313.5	$328.5	$338.1	$356.8	$375.0	$399.3

[a] Includes automobiles and parts, furniture and household equipment, and other durable goods.

[b] Includes food and beverages, clothing and shoes, gasoline and oil, and other nondurable goods.

[c] Includes housing, household operation, transportation, and other services.

[d] Details may not add to total due to rounding.

SOURCE: U.S. Department of Commerce, Office of Business Economics, *Survey of Current Business*, p. S-1.

7. Construct a semilogarithmic line diagram using the data in Table 2.6.

TABLE 2.6 POPULATION OF THE UNITED STATES, CALIFORNIA, AND LOS ANGELES, 1900–60

(Millions)

Year	United States	California	Los Angeles
1900	76.0	1.5	0.1
1910	92.0	2.4	0.3
1920	105.7	3.5	0.6
1930	122.8	5.7	1.2
1940	131.7	6.9	1.5
1950	150.7	10.6	2.0
1960	178.5	15.8	2.5

SOURCE: U.S. Department of Commerce, Bureau of the Census, *Decennial Census of Population.*

3

Frequency Distributions

This chapter reviews the technique for classification of large quantities of data in order to prepare them for further analysis. When hundreds and even thousands of numbers are involved, some type of format is required in order to present these data in some compact form. The device used by the statistician is known as the frequency distribution, *a quantitative classification of statistical data.*

The chapter presents a sample of weekly earnings of 100 clerks, illustrating the arrangement of data in size order, known as an array *and the application of the* array.

The steps in grouping data into a frequency distribution follow. The chapter then discusses the advantages of grouping data, and the types of graphic presentation available for visualizing a frequency distribution, namely, the frequency polygon and histogram.

The normal curve is introduced graphically, and its characteristics are given. In addition, asymmetrical curves are illustrated as well as the graph of a bimodal distribution and the J-shaped curve.

The discussion of the frequency distribution begins the analysis of descriptive statistics. *By* descriptive statistics (*or* descriptive statistical methods) *is meant analysis of statistical data through rearrangement in a more useful form and the calculation of such basic measures as averages and measures of variation. The origin of descriptive statistics dates from Biblical times. In the Book of Numbers* (1:1–3) *Moses was directed to conduct a population census among the Hebrews who had been freed from Egyptian bondage. This is probably the first recorded census.*

3.1 INTRODUCTION TO FREQUENCY DISTRIBUTIONS

The trade association of a particular industry may wish to analyze clerical worker salaries in a major city. To collect weekly earnings data for every clerical worker would be a mammoth task. The association's statisticians may decide therefore to draw a sample of weekly earnings of 100 clerks from among all the firms in the industry of that city. Then, by studying the sample, it becomes possible to make some generalizations about wages in the industry.

To illustrate the frequency distribution and its method of construction, a sample of 100 clerical workers' salaries is shown in Table 3.1. These data will be grouped in various ways in order ultimately to derive the frequency distribution.

Each numerical value in Table 3.1 is referred to as an *observation* or an *item*.

TABLE 3.1 WEEKLY EARNINGS OF A SAMPLE OF 100 CLERKS, 1960

$ 76.75	$ 89.87	$ 95.37	$ 62.55	$ 72.65
87.60	108.47	100.00	72.60	88.80
85.00	63.83	70.01	68.30	68.22
82.54	100.50	80.00	110.06	74.32
90.00	55.10	60.11	54.00	62.73
86.00	69.20	89.88	68.09	65.78
59.00	67.01	90.40	80.00	71.00
85.40	101.89	72.00	64.88	62.60
110.42	64.40	58.00	54.00	67.65
95.00	79.11	63.65	56.40	61.00
52.80	75.00	55.00	62.00	57.70
93.20	73.10	75.10	119.00	55.00
61.00	125.82	64.00	65.58	66.00
67.35	53.60	79.00	98.55	127.27
80.00	55.00	90.01	74.09	71.55
97.45	71.00	90.50	65.78	68.53
56.16	73.17	60.94	73.66	100.12
80.64	105.50	63.82	76.00	65.80
76.55	64.93	70.09	75.00	87.79
66.55	70.84	51.67	71.75	81.60

SOURCE: Hypothetical data.

3.2 QUANTITATIVE CLASSIFICATION OF DATA

Ungrouped Data

An elementary type of analysis involves arranging ungrouped data in size order, i.e., as an *array*. The smallest value is first and the largest last.

Subsequently, this arrangement of individual figures may serve as the first step in grouping the figures.

DEFINITION: The *array* is an arrangement of data according to their magnitude. It is illustrated in Table 3.2.

TABLE 3.2 ARRAY OF WEEKLY EARNINGS OF A SAMPLE OF 100 CLERKS, 1960

$51.67	$62.60	$68.30	$75.10	$90.00
52.80	62.73	68.53	76.00	90.01
53.60	63.65	69.20	76.55	90.40
54.00	63.82	70.01	76.75	90.50
54.00	63.83	70.09	79.00	93.20
55.00	64.00	70.84	79.11	95.00
55.00	64.40	71.00	80.00	95.37
55.00	64.88	71.00	80.00	97.45
55.10	64.93	71.55	80.00	98.55
56.16	65.58	71.75	80.64	100.00
56.40	65.78	72.00	81.60	100.12
57.70	65.78	72.60	82.54	100.50
58.00	65.80	72.65	85.00	101.89
59.00	66.00	73.10	85.40	105.50
60.11	66.55	73.17	86.00	108.47
60.94	67.01	73.66	87.60	110.06
61.00	67.35	74.09	87.79	110.42
61.00	67.65	74.32	88.80	119.00
62.00	68.09	75.00	89.87	125.82
62.55	68.22	75.00	89.88	127.27

SOURCE: Table 3.1.

The sum of the 100 items shown in the table is $7,636.37.

APPLICATION: The *array* is a useful tool to statisticans in the following ways:

1. The array shows clearly the extreme values and the range between them.
2. It indicates the values around which the items tend to cluster.
3. The array is essential for deriving certain statistical measures, considered in Chapter 4.

Grouped Data

The ungrouped data may now be grouped into a frequency distribution.

DEFINITIONS: A *frequency distribution* is an orderly arrangement of a series of data grouped into classes which are listed according to magnitude.

A *class* is a grouping of data within two boundary values, a lower class limit and an upper class limit.

A *class limit* is a boundary value of a class. For example, \$50.00 to \$59.99 is a class whose lower limit is \$50.00 and whose upper limit is \$59.99.

A *class interval* is a value indicating the difference between a *lower class limit* and an *upper class limit*. For example, for the class \$50.00 to \$59.99, the class interval is \$10 (\$59.99 − \$50.00 = \$9.99, which for convenience is rounded to \$10).

Steps in Grouping Data or Constructing a Frequency Distribution

STEP 1. Determine the *size of the class interval:* The size of the class interval determines the number of classes in the frequency distribution, or the number of groups into which the observations are divided. A frequency distribution may have more classes or fewer classes, or wider class intervals or narrower class intervals. As a rule of thumb, the number of classes should not be fewer than 5 nor more than 15.

A formula useful in approximating the size of the class interval was developed by H. A. Sturges in 1926:

$$i = \frac{\text{range}}{1 + 3.322 \log n}$$

where i = class interval

Range = difference between highest and lowest value in series

n = number of observations in sample, for which log (logarithm to the base 10) is entered in formula.[1]

EXAMPLE: The class interval size for the weekly earnings of 100 clerks in 1960 may be determined as follows:

$$\begin{aligned} i &= \frac{\$127.27 - \$51.67}{1 + (3.322)(\log 100)} \\ &= \frac{\$75.60}{1 + (3.322)(2.0000)} \\ &= \frac{\$75.60}{1 + 6.644} \\ &= \frac{\$75.60}{7.644} \\ &= \$9.89 \end{aligned}$$

For convenience, the suggested \$9.89 class interval is rounded to \$10.00.

[1] See Appendix A, How to Use Logarithms.

STEP 2. Set up the class intervals, and tabulate the data.

Instead of starting the first class with the lowest value in the series, $51.67, a lower class limit of $50.00 is used in order to provide a neater set of class limits.

Each class is mutually exclusive as a result of setting the upper class limit of each class at one cent less than the ten dollar interval, or $50.00–$59.99, rather than $50.00–$60.00. If the latter class limits were used, a problem would arise if weekly earnings of $60.00 appeared in the sample. It would be impossible to decide to allocate it to the class $50.00–$60.00 or $60.00–$70.00.

In actual practice, the array is seldom prepared for large distributions. Ungrouped data are usually tallied into a frequency distribution after the limits of the classes have been set. This procedure is shown in Table 3.3 using the data of Table 3.1.

TABLE 3.3 TALLY SHEET FOR GROUPING WEEKLY EARNINGS OF 100 CLERKS, 1960, INTO FREQUENCY DISTRIBUTION OF 8 CLASSES, EACH $10.00 WIDE, WITHOUT USE OF ARRAY

Weekly earnings (8 classes)	Number of clerks
$50.00–$59.99	卌 卌 卌
60.00– 69.99	卌 卌 卌 卌 卌 ||||
70.00– 79.99	卌 卌 卌 卌 |||
80.00– 89.99	卌 卌 ||||
90.00– 99.99	卌 ||||
100.00–109.99	卌 |
110.00–119.99	|||
120.00–129.99	||

SOURCE: Table 3.1.

Sometimes the grouping into a frequency distribution is made by means of an entry form. This procedure is shown in Table 3.4, using the data of Table 3.1.

In Table 3.4 each observation given in Table 3.1 is actually transcribed in its appropriate class. While the procedure of tallying the entries, as in Table 3.3, is simpler, Table 3.4 is easier to recheck if an error is discovered.

STEP 3. Make the frequency distribution.

The frequency distribution is derived from Tables 3.3 or 3.4, whichever is used. Table 3.5 presents the frequency distribution in summary form.

TABLE 3.4 ENTRY FORM FOR GROUPING WEEKLY EARNINGS OF 100 CLERKS, 1960, INTO FREQUENCY DISTRIBUTION OF 8 CLASSES, EACH $10.00 WIDE

$50.00–59.99	$60.00–69.99	$70.00–79.99	$80.00–89.99	$90.00–99.99	$100.00–109.99	$110.00–119.99	$120.00–129.99
$59.00	$61.00	$76.75	$87.60	$90.00	$108.47	$110.42	$125.82
52.80	67.35	76.55	85.00	95.00	100.50	110.06	127.27
56.16	66.55	79.11	82.54	93.20	101.89	119.00	
55.10	63.83	75.00	86.00	97.45	105.50		
53.60	69.20	73.10	85.40	95.37	100.00		
55.00	67.01	71.00	80.00	90.40	100.12		
58.00	64.40	73.17	80.64	90.01			
55.00	64.93	70.84	89.87	90.50			
51.67	60.11	70.01	80.00	98.55			
54.00	63.65	72.00	89.88				
54.00	64.00	75.10	80.00				
56.40	60.94	79.00	88.80				
57.70	63.82	70.09	87.79				
55.00	62.55	72.60	81.60				
	68.30	74.09					
	68.09	73.66					
	64.88	76.00					
	62.00	75.00					
	65.58	71.77					
	65.78	72.65					
	68.22	74.32					
	62.73	71.00					
	65.78	71.55					
	62.60						
	67.65						
	61.00						
	66.00						
	68.53						
	65.80						

SOURCE: Table 3.1.

STEP 4. Draw a graph of the frequency distribution.

The primary reason for drawing a graph of a frequency distribution is to present the data visually as an aid in analysis. The two types of graphs most widely used for this purpose are the arithmetic line diagram and the bar chart which are called the *frequency polygon* and *histogram*, respectively, when used to graph a frequency distribution.

TABLE 3.5 FREQUENCY DISTRIBUTION OF WEEKLY EARNINGS OF 100 CLERKS, 1960

Weekly earnings	Number of clerks (frequency)
$50.00–$59.99	14
60.00– 69.99	29
70.00– 79.99	23
80.00– 89.99	14
90.00– 99.99	9
100.00–109.99	6
110.00–119.99	3
120.00–129.99	2
	n = 100

SOURCE: Tables 3.3 or 3.4.

DEFINITIONS: The *frequency polygon* (Figure 3.1) is a graphic device for presenting a frequency distribution. It consists of a series of connected line segments. Each line segment joins the points representing the frequency (shown on the Y scale) for each class interval (shown on the X scale).

The *histogram* (Figure 3.2) is a graphic device for presenting a frequency distribution. The bars are drawn so that the height of each corresponds to

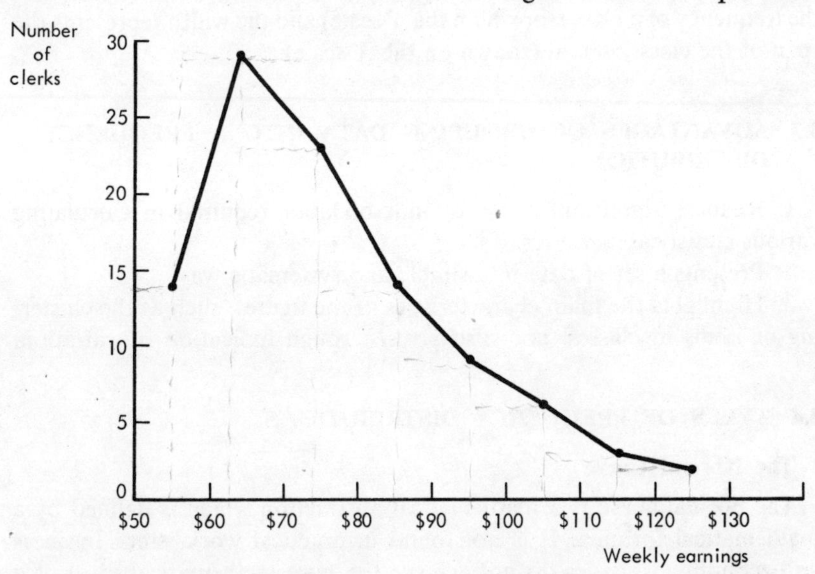

Fig. 3.1 A Frequency Polygon for Weekly Earnings of 100 Clerks, 1960
SOURCE: Table 3.5.

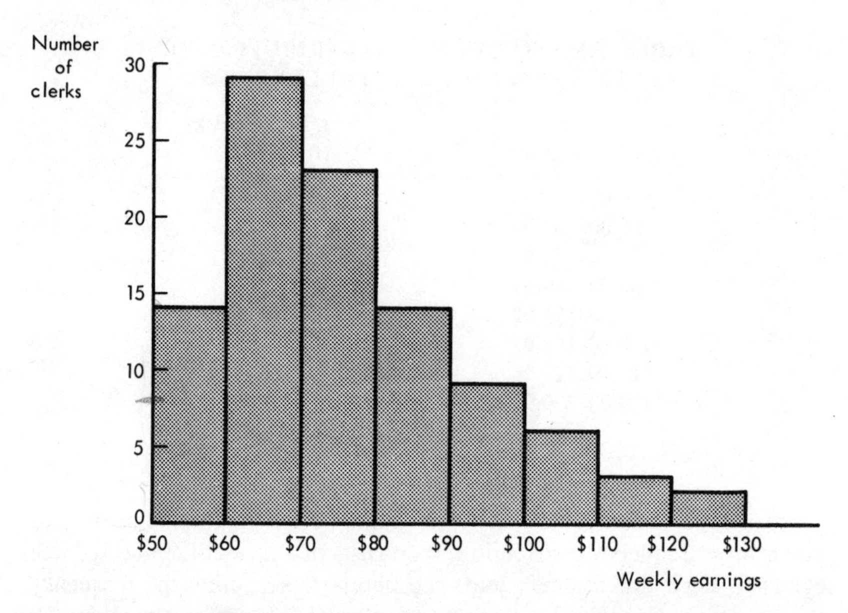

Fig. 3.2 Histogram for Weekly Earnings of 100 Clerks, 1960
SOURCE: Table 3.5.

the frequency of a class (shown on the *Y* scale) and the width represents the span of the class interval (shown on the *X* scale).

3.3 ADVANTAGES OF GROUPING DATA INTO A FREQUENCY DISTRIBUTION

1. Reduces significantly the amount of labor required in calculating various statistical measures.
2. Presents a set of data in a simple and systematic way.
3. Highlights the main characteristics of the figures, such as the clustering of items in classes, and also gives a rough indication of variation.

3.4 TYPES OF FREQUENCY DISTRIBUTIONS

The Normal Curve

The normal curve is a mathematical abstraction which is defined by a mathematical formula. It is not found in practical work, since business and economic statistics do not achieve the pure mathematical ideal of a perfect frequency distribution. The normal curve is useful, however, because it permits the statistician to measure the extent to which real data

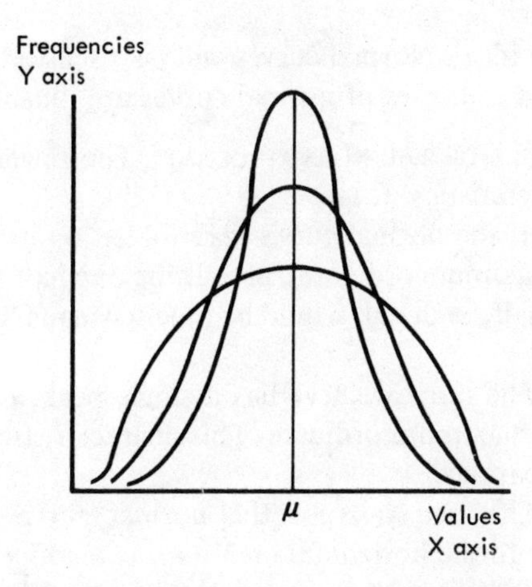

Fig. 3.3 Examples of Normal Curves

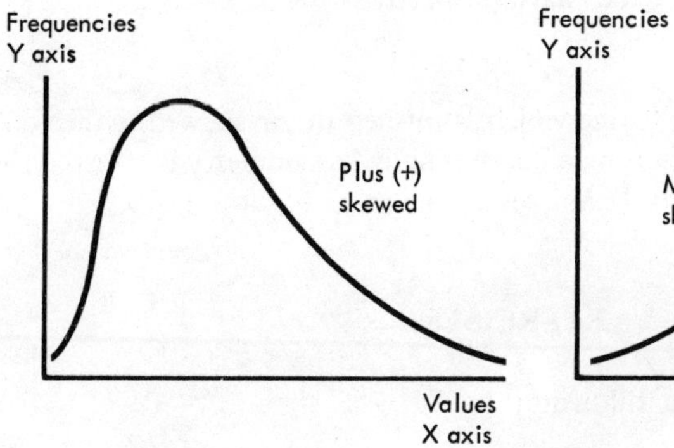

Fig. 3.4 Right-skewed, or Positively Skewed, Frequency Curve

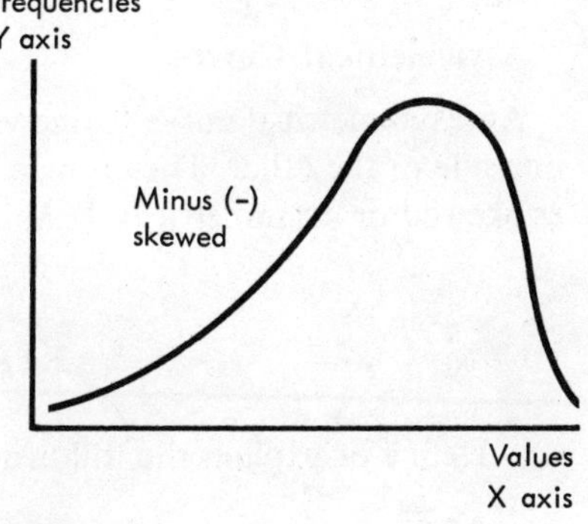

Fig. 3.5 Left-skewed, or Negatively Skewed, Frequency Curve

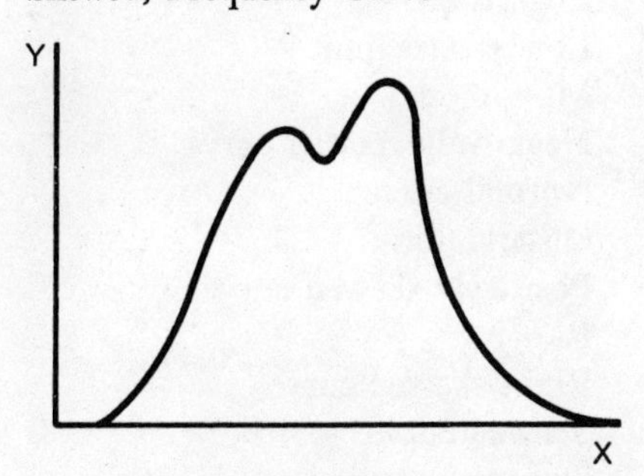

Fig. 3.6 Bimodal Distribution: Frequency Curve with Two Peaks, or Modes

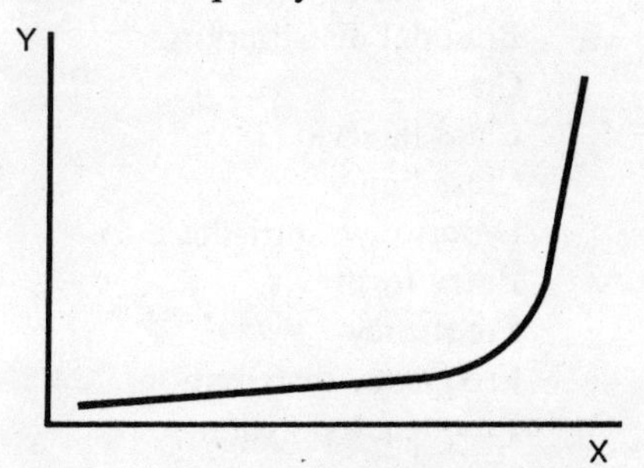

Fig. 3.7 J-shaped Curve, or J-Curve, Which Resembles the Capital Letter J

deviate from the ideal. Normal curves will be discussed in greater detail in Chapter 6. Three *examples* of normal curves are illustrated in Figure 3.3.

CHARACTERISTICS OF THE NORMAL CURVE: The normal curve has three important characteristics. It is

Symmetrical: If the normal curve were folded on its vertical center line (known as the maximum ordinate), permitting one half to be superimposed upon the other half, each half would be precisely equal to the other in both size and shape.

Monomodal: The normal curve has a single peak, or mode, located at the point of the maximum ordinate. This characteristic gives the curve its bell-shaped appearance.

Asymptotic: The two tails of the normal curve continue outward closer and closer to the horizontal or X axis, extending out to infinity, but never touch the X axis. The normal curve is therefore described as approaching the horizontal axis asymptotically.

Asymmetrical Curves

An asymmetrical curve is one which is pushed in, or skewed, either on one side or the other. Therefore, a curve that lacks symmetry is referred to as skewed or asymmetrical. Examples are given in Figures 3.4–3.7.

EXERCISES

1. Define or explain the following:

Array
Asymmetrical curve
Asymptotic
Bimodal distribution
Class
Class interval
Class limit
Descriptive statistics
Entry form
Frequency
Frequency distribution
Frequency polygon
Histogram
Item
J-shaped curve
Left-skewed curve
Logarithm
Lower class limit
Monomodal
Negatively skewed curve
Normal curve
Observation
Positively skewed curve
Range
Right-skewed curve
Skewed curve
Sturges formula
Symmetrical
Tally sheet

2. Given the data in Table 3.6:

TABLE 3.6 GROSS MONTHLY RENTS OF A SAMPLE OF 75 RENT-CONTROLLED APARTMENTS, NEW YORK CITY, 1964

$32.50	$74.75	$32.00
32.25	65.39	42.00
56.20	59.00	55.90
48.00	72.50	36.31
91.25	45.00	47.80
85.00	72.00	47.30
78.00	58.00	23.15
100.00	45.00	48.00
115.00	66.00	22.48
115.00	129.00	48.00
60.00	65.00	90.30
35.00	80.00	86.00
80.00	103.35	50.00
106.00	40.00	59.15
71.50	25.12	80.00
71.00	38.95	64.85
28.30	36.95	34.85
41.07	27.25	55.00
51.16	43.60	40.00
54.22	74.00	53.50
75.00	90.00	42.32
25.00	57.50	47.50
52.04	25.00	75.50
27.00	81.50	28.22
52.00	23.00	25.00

a. Arrange the items in an array.
b. Estimate the size of the class interval for a frequency distribution by using the Sturges formula.
c. Construct a frequency distribution using a class interval of $20, with $20.00 to $39.99 as the lower and upper limits for the first class, $40.00 to $59.99 for the second, etc.
d. On a sheet of arithmetic graph paper, prepare a frequency polygon of these data.
e. On a sheet of arithmetic graph paper, prepare a histogram of these data.

3. Answer the following questions relating to the frequency distribution constructed in answer to Exercise 2c above.

   a. What is the value of n?
   b. What is the value of i?
   c What is the midpoint of the first interval?
   d. What is the lower limit of the first interval?
   e. What is the upper limit of the last interval?
   f. How many classes are there in the frequency distribution?

4. Table 3.7 presents grades on an examination question worth 20 points. Construct a frequency distribution of about 10 classes.

TABLE 3.7 GRADES ON AN EXAMINATION QUESTION WORTH 20 POINTS

Grade	Number of students	Grade	Number of students
14.3	5	15.8	18
14.4	6	15.9	16
14.5	4	16.0	14
14.6	9	16.1	13
14.7	7	16.2	11
14.9	8	16.3	10
15.0	11	16.4	8
15.1	12	16.5	6
15.2	14	16.6	8
15.3	13	16.7	7
15.4	15	16.8	6
15.5	17	16.9	8
15.6	16	17.0	4
15.7	12	17.1	5

4

Measures of Location

Statistical analysis seeks to develop concise summary figures which describe a large body of quantitative data. One of the most widely used set of summary figures is known as measures of location, which are often referred to as averages, measures of central tendency, or central location. The purpose for computing an average value for a set of obesrvations is to obtain a single value which is representative of all the items, and which the mind can grasp simply and quickly. This single value is the point or location around which the individual items cluster.

The layman is familiar with many examples of averages, such as average weekly earnings, average income, a batting average, or average grades. Each of these examples is somewhat different in concept, and in statistics a different procedure is required to compute each type of average.

Four kinds of measures of location are analyzed in this chapter:

1. *The arithmetic mean*
2. *The mode*
3. *The geometric mean*
4. *The partition values, of which the median is the most important*

The analysis for each measure includes a definition, an example with the solution, the formula, the characteristics of the measure, and the advantages and disadvantages of each.

4.1 THE ARITHMETIC MEAN

The arithmetic mean is the most commonly used measure of location. The principles and methods which follow distinguish between grouped data and ungrouped data.

DEFINITIONS:

Ungrouped data are values which are not organized or structured in any way.

Grouped data are values which are grouped into classes with specified lower and upper limits for each class.

Calculating the Arithmetic Mean for Ungrouped Data

DEFINITION: The *arithmetic mean* of a series of observations is the sum of the observations divided by the total number of observations in the series.

EXAMPLE: The arithmetic mean of the weekly earnings of the sample of 100 clerks is computed by obtaining the sum of the 100 observations and then dividing that sum by 100.

$$\frac{\$7{,}636.37}{100} = \$76.36$$

FORMULA:

$$\bar{x} = \frac{\Sigma x}{n}$$

where $\bar{x}$ = arithmetic mean

Σ = Greek capital letter *sigma*, used as the symbol for "the sum of"

x = individual observations or items

n = number of items

Calculating the Arithmetic Mean for Grouped Data: The Long Method

To calculate the arithmetic mean from observations arranged as a frequency distribution, it is necessary to use the classes of the distribution and their midpoints.

DEFINITION: The *midpoint* of each class is approximately equal to the arithmetic average of the items which fall in that class. The midpoint of the first class is \$54.995, derived from the following:

$$\frac{\$50.00 + \$59.99}{2} = \$54.995, \quad \text{or} \quad \$55.00.$$

EXAMPLE: What is the arithmetic mean of weekly earnings of the 100 clerks shown in Table 3.5?

TABLE 4.1 CALCULATION OF ARITHMETIC MEAN BY THE LONG METHOD FOR GROUPED DATA

Weekly earnings (l_1–l_2)	Number of clerks (f)	Class midpoint (x)	Total earnings (fx)
$50.00–$59.99	14	$ 55	$ 770
60.00– 69.99	29	65	1,885
70.00– 79.99	23	75	1,725
80.00– 89.99	14	85	1,190
90.00– 99.99	9	95	855
100.00–109.99	6	105	630
110.00–119.99	3	115	345
120.00–129.99	2	125	250
	n = 100		Σfx = $7,650

SOURCE: Table 3.5.

Formula:

$$\bar{x} = \frac{\Sigma fx}{n}$$

$$= \frac{\$7,650}{100}$$

$$= \$76.50$$

In Table 4.1, l_1 and l_2 in the first column indicate the lower limit (l_1) and the upper limit (l_2) of each class. The new term in the formula is f, which is the frequency, or number of items in each class.

Calculating the Arithmetic Mean for Grouped Data: The Short Method

The short method for computing the mean seeks to reduce the amount of arithmetic involved. The method requires that an arithmetic mean be guessed or assumed, and then after using the guessed mean in the computations, a correction is made for the amount of error in the guess.

The guessed mean ($\bar{x}_g$) may be the midpoint of any class. For the class selected, the value "zero" (0) is assigned in the d or deviation column, and the classes above and below it are counted as indicated in Table 4.2.

The new element introduced in Table 4.2 is the d column, which is the class deviation from the guessed mean class. This is measured −1, −2, −3, etc., above the assumed mean class, and +1, +2, +3, etc., below the assumed mean class.

TABLE 4.2 CALCULATION OF ARITHMETIC MEAN BY THE SHORT METHOD FOR GROUPED DATA

Weekly earnings (l_1–l_2)	Number of clerks (f)	Class midpoint (x)	Class deviation (d)	fd	
\$50.00–\$59.99	14	\$ 55	−2	−28	−57
60.00– 69.99	29	65	−1	−29	
70.00– 79.99	23	75 = $\bar{x}_g$	0	0	
80.00– 89.99	14	85	+1	+14	+72
90.00– 99.99	9	95	+2	+18	
100.00–109.99	6	105	+3	+18	
110.00–119.99	3	115	+4	+12	
120.00–129.99	2	125	+5	+10	
	n = 100			Σfd = +15	

Formula: $\bar{x} = \bar{x}_g + \left(\frac{\Sigma fd}{n}\right) i$

where $\bar{x}_g$ = guessed mean

Σfd = sum of the frequency times the class deviation

n = number of items

i = class interval

Solution: $\bar{x} = \$75 + \left(\frac{15}{100}\right) (\$10)$

$= \$75 + \frac{150}{100}$

$= \$75 + \$1.50$

$= \$76.50$

SOURCE: Table 3.5.

CHECK OF COMPUTATIONS (Table 4.3): To prove that any midpoint may be selected as the guessed mean ($\bar{x}_g$) and a correct arithmetic mean would be obtained, another class midpoint is selected. The identical arithmetic mean is derived.

TABLE 4.3 CALCULATION OF ARITHMETIC MEAN BY THE SHORT METHOD FOR GROUPED DATA: PROOF

Weekly earnings (l_1–l_2)	Number of clerks (f)	Class midpoint (x)	Class deviation (d)	fd	
\$50.00–\$59.99	14	\$ 55	−4	−56	−203
60.00– 69.99	29	65	−3	−87	
70.00– 79.99	23	75	−2	−46	
80.00– 89.99	14	85	−1	−14	
90.00– 99.99	9	95 = $\bar{x}_g$	0	0	
100.00–109.99	6	105	+1	+6	+18
110.00–119.99	3	115	+2	+6	
120.00–129.99	2	125	+3	+6	
	n = 100			Σfd = −185	

$$\textit{Formula: } \bar{x} = \bar{x}_g + \left(\frac{\Sigma fd}{n}\right) i$$

$$\textit{Solution: } \bar{x} = \$95 + \left(\frac{-185}{100}\right)(10)$$

$$\bar{x} = \$95 - (1.85)(10)$$

$$\bar{x} = \$95 - \$18.50$$

$$\bar{x} = \$76.50$$

ANALYSIS: If the guessed mean is too low, the correction factor ($\Sigma fd/n$) will be positive, as indicated in the previous solution. If the guessed mean is too high, the correction factor ($\Sigma fd/n$) will be negative, as indicated above. In either case, the identical arithmetic mean is obtained.

CONCLUSION: The mean found by the short method for grouped data is identical with the mean computed by the long method for grouped data. It differs, however, from the mean computed for ungrouped data.

The figures used in the short method are simpler than in the long method. The computations in the short method, therefore, are easier than in the long method.

4.2 CHARACTERISTICS OF THE ARITHMETIC MEAN

1. The arithmetic mean is influenced by each and every item in a distribution. Therefore, extreme values have a significant influence on the value of the mean.

2. The algebraic sum of the deviations of each value from the arithmetic mean of the series is always equal to zero.

3. The sum of the squared deviations of each value from the arithmetic mean of the series is always less than the sum of the squared deviations computed from either the median or the mode.

4. By weighting the means of two or more frequency distributions, it is possible to develop a mean for the combined groups.

4.3 ADVANTAGES OF THE ARITHMETIC MEAN

1. The arithmetic mean is a simple, easily understood measure which gives due weight to every value in the distribution.
2. It may be computed from original data without preparing an array or frequency distribution.
3. It lends itself to subsequent algebraic treatment.
4. The arithmetic mean of a sample will be closer to the true mean of the population than the median of a sample.

4.4 DISADVANTAGES OF THE ARITHMETIC MEAN

1. The arithmetic mean may be influenced too much by extreme values.
2. The arithmetic mean should be avoided if the data are badly skewed.

4.5 THE MEDIAN

DEFINITION: The *median* is a position average, a value so located in the frequency distribution that it divides it in half, with 50 per cent of the items below it and 50 per cent of the items above it.

Calculating the Median for Ungrouped Data

To compute the median from ungrouped data, it is necessary first to make an array.

AN EVEN NUMBER OF ITEMS: If the number of items for which the median is to be computed is even, the following formula is used.

FORMULA:

$$\text{Median} = \frac{X_{1(n/2)} + X_{2[(n+2)/2]}}{2}$$

where

$$X_{1(n/2)} = \text{value for item } \frac{n}{2}$$

$$X_{2[(n+2)/2]} = \text{value for item } \frac{n+2}{2}$$

Example: To compute the median for weekly earnings of 100 clerks, the array of Table 3.2 is used. The two middle items are selected, the fiftieth and fifty first entries, and the mean of the two values is computed.

Solution:

$$X_{1(n/2)} = \frac{100}{2} = \text{50th entry} = \$71.75$$

$$X_{2[(n+2)/2]} = \frac{100+2}{2} = \frac{102}{2} = \text{51st entry} = \$72.00$$

$$\text{Median} = \frac{\$71.75 + \$72.00}{2}$$

$$= \frac{\$143.75}{2}$$

$$= \$71.875, \quad \text{or} \quad \$71.88$$

AN ODD NUMBER OF ITEMS: For a distribution with an odd number of items, the median value is the middle item of the array, or the value equivalent to the item number.

$$X_{[(n+1)/2]}$$

Calculating the Median for Grouped Data (Table 4.4)

To compute the median from grouped data, an interpolation procedure is used.

EXAMPLE: Given the frequency distribution of Table 3.5, compute the median.

TABLE 4.4 CALCULATION OF THE MEDIAN FOR GROUPED DATA

Weekly earnings (l_1–l_2)	Number of clerks (f)	
\$50.00–\$59.99	14	43 = cumulated f counting down (14 + 29)
60.00– 69.99	29	
70.00– 79.99	23	
80.00– 89.99	14	34 = cumulated f counting up (14 + 9 + 6 + 3 + 2)
90.00– 99.99	9	
100.00–109.99	6	
110.00–119.99	3	
120.00–129.99	2	
	n = 100	

SOURCE: Table 3.5.

Steps for computation:

1. Find $n/2$, retaining the decimal if any.
2. Count down (or count up) in the frequency column to the median class, and subtract the accumulated frequencies from $n/2$.
3. Divide the residual thus found by the frequency of the median class, and multiply by the class interval.
4. Add the result to the lower limit of the median class, if counting down; subtract it from the upper limit, if counting up.

Computation—counting down:

1. $\frac{n}{2} = \frac{100}{2} = 50$
2. $50 - 43 = 7$
3. $\frac{7}{23}(\$10) = \frac{70}{23} = \$3.0434$
4. $\$70.00 + \$3.04 = \$73.04$

Computation—counting up:

This is a check on the computation results obtained by counting down. The steps for computation are the same.

1. $\frac{n}{2} = \frac{100}{2} = 50$
2. $50 - 34 = 16$
3. $\frac{16}{23}(\$10) = \frac{160}{23} = \$6.956$
4. $\$79.999 - \$6.956 = \$73.04$

FORMULA FOR COUNTING DOWN:

$$\text{Median} = l_1 + \left(\frac{\frac{n}{2} - \Sigma f_c}{f_{\text{med}}}\right) i$$

where l_1 = lower limit of class containing median item
n = number of items in distribution
Σ = sum of
f_c = cumulated frequencies
f_{med} = frequency of median class
i = class interval

FORMULA FOR COUNTING UP:

$$\text{Median} = l_2 - \left(\frac{\frac{n}{2} - \Sigma f_c}{f_{\text{med}}}\right) i$$

where l_2 = upper limit of class containing median item

4.6 CHARACTERISTICS OF THE MEDIAN

1. The median, an average of position, is not affected by extreme values. Because of this, the median is more widely used than the arithmetic mean as a measure of central tendency for income distributions. Very large incomes would have no influence on the value of the median.
2. To compute the median it is necessary to arrange the observations in size order either as ungrouped or grouped data.

4.7 ADVANTAGES OF THE MEDIAN

1. The median is easy to understand and compute.
2. It may be used when a distribution has extreme values which would distort the arithmetic mean.
3. It may be computed in an open-end distribution, one in which the lowest-value class has no lower limit and the highest-value class has no upper limit.
4. It is the best available measure for deriving a typical value in an income distribution.
5. It can be computed from the graph of the intersection of the "more than" and "less than" ogives (see Section 4.9).

4.8 DISADVANTAGES OF THE MEDIAN

1. The median is not as widely known as the arithmetic mean.
2. The data must be arranged according to size (arrayed) in order to calculate the median.
3. The medians of several series cannot be averaged to yield a median of all the series combined.
4. It is not too useful if the data do not cluster closely at the distribution's center.

4.9 THE OGIVE OR CUMULATIVE FREQUENCY CURVE

DEFINITION: The *ogive* is the graphic presentation of cumulated frequencies of a frequency distribution.

EXAMPLE: Table 4.5 illustrates the "less than" and the "more than" cumulation of frequencies.

TABLE 4.5 UNCUMULATED AND CUMULATED FREQUENCIES

Weekly earnings (l_1–l_2)	Number of clerks (f)	Lower limits of classes	Number: "Less than" the lower limit	Number: "More than" the lower limit
$50.00–$59.99	14	$ 50	0	100
60.00– 69.99	29	60	14	86
70.00– 79.99	23	70	43	57
80.00– 89.99	14	80	66	34
90.00– 99.99	9	90	80	20
100.00–109.99	6	100	89	11
110.00–119.99	3	110	95	5
120.00–129.99	2	120	98	2
	n = 100	130	100	0

SOURCE: Table 3.5.

Setting Up the Cumulative Frequency Table (Table 4.5)

1. The first entry in the column Lower Limits of Classes is the lower limit (l_1) of the first class in the frequency distribution. Only the single value is used.

2. The last entry in the Lower Limits column is the upper limit (l_2) of the last class in the frequency distribution. This value, $129.99, is rounded to $130.

3. The column Less Than the Lower Limit cumulates frequencies downward, and the ogive drawn from these data is called the "less than"ogive.

4. The column More Than the Lower Limit cumulates frequencies upward, and the ogive drawn from these data is called the "more than" ogive.

Transformation of Cumulated Frequencies into Cumulated Percentages

Usually, after the frequencies of the ogive are cumulated, they are transformed from actual numbers into cumulative percentages. These percentages are plotted on the Y axis, rather than the actual frequencies, so that the per cent "more than" or "less than" a particular value can be read directly from the graph.

Graphic Presentation of the Ogive

Figure 4.1 presents the "less than" and "more than" ogives. If the two ogives are plotted accurately on the same graph, they will intersect at a point above the X axis from which the *median* can be determined.

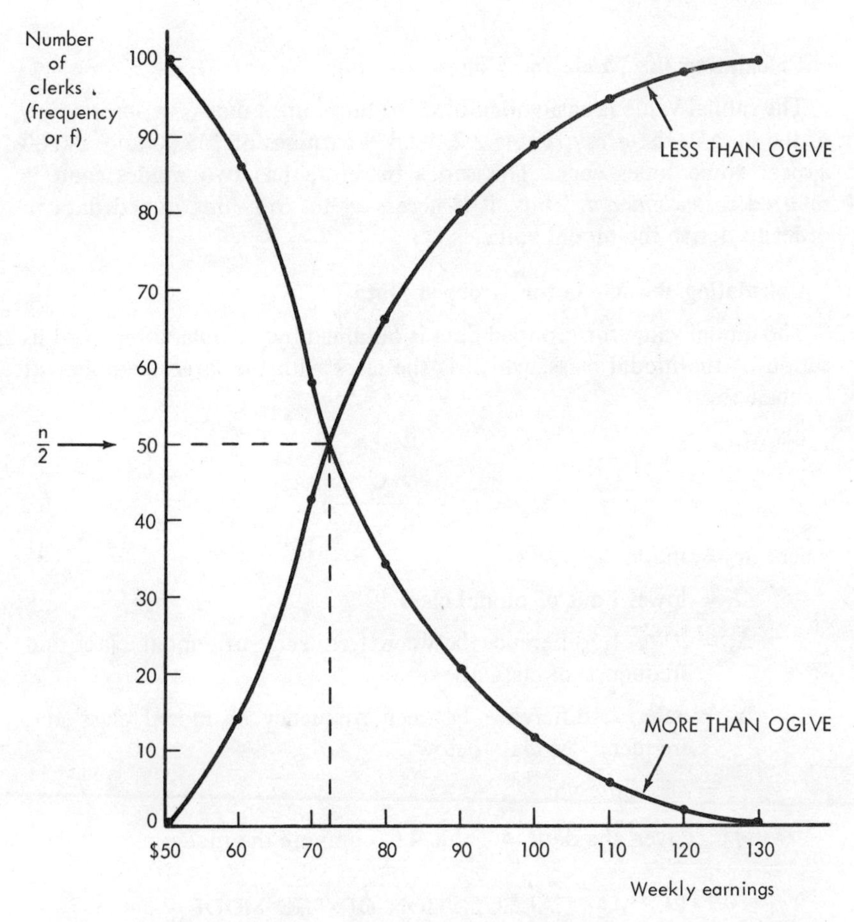

Fig. 4.1 Weekly Earnings of 100 Clerks, 1960: "More Than" and "Less Than" Ogives

SOURCE: Table 4.5.

4.10 LOCATING THE MEDIAN GRAPHICALLY FROM THE OGIVE

To locate the median graphically from the ogive, drop a perpendicular from the intersection of the "more than" and "less than" ogive to the horizontal, or X axis. The median value may be read from the X axis. In this case, the median graphically is about $73.00.

4.11 THE MODE

DEFINITION: The *mode* is the value of a series that appears most frequently. This value is called the modal value.

Calculating the Mode for Ungrouped Data

The modal value is easily identified for ungrouped data. By observation of the data in the array, Table 3.2, weekly earnings of $55.00 and $80.00 appear three times each. The series therefore has two modes, and is referred to as *bimodal*. Thus, it is necessary to array ungrouped data in order to derive the modal value.

Calculating the Mode for Grouped Data

The modal value for grouped data is obtained by formula, after identification of the modal class, which is the class with the largest number of frequencies.

FORMULA:

$$M_o = l_1 + \frac{\Delta_1}{\Delta_1 + \Delta_2} i$$

where M_0 = mode

l_1 = lower limit of modal class

Δ_1 = *delta* 1, difference between frequency of modal class and frequency of class above

Δ_2 = *delta* 2, difference between frequency of modal class and frequency of class below

i = class interval

EXAMPLE: Given the data in Table 4.6, compute the mode.

TABLE 4.6 CALCULATION OF THE MODE

Weekly earnings (l_1–l_2)	Number of clerks (f)
$50.00–$59.99	14
60.00– 69.99	29 = Modal class
70.00– 79.99	23
80.00– 89.99	14
90.00– 99.99	9
100.00–109.99	6
110.00–119.99	3
120.00–129.99	2
	n = 100

SOURCE: Table 3.5.

Solution: The mode for weekly earnings of 100 clerks is as follows:

$$
\begin{aligned}
M_o &= \$60.00 + \frac{15}{15 + 6}(\$10) \\
&= \$60.00 + \frac{15}{21}(\$10) \\
&= \$60.00 + \frac{150}{21} \\
&= \$60.00 + \$7.14 \\
&= \$67.14
\end{aligned}
$$

4.12 CHARACTERISTICS OF THE MODE

1. The mode represents the value which is most typical or occurs most often in a distribution. As such, it is affected neither by extreme values (as is the arithmetic mean) nor by the number and position of other values (as is the median).
2. The value of the mode is affected significantly by the size of the class interval used in grouping data into a frequency distribution. A change in the size of the class interval will change the value of the mode.

4.13 ADVANTAGES OF THE MODE

1. It is the most typical value, and therefore most descriptive.
2. It can be computed in open-end distributions because it is not influenced by either the number or value of items in the tails of a distribution.
3. It is simple to compute for grouped data and from the array of ungrouped data.
4. It can be computed from the graph of the histogram.

4.14 DISADVANTAGES OF THE MODE

1. In a series of ungrouped data, the mode does not exist if each value appears only once.
2. It is affected by pre-established class limits and class intervals.
3. A series may have two or more modes, thus diminishing its usefulness as a typical value.

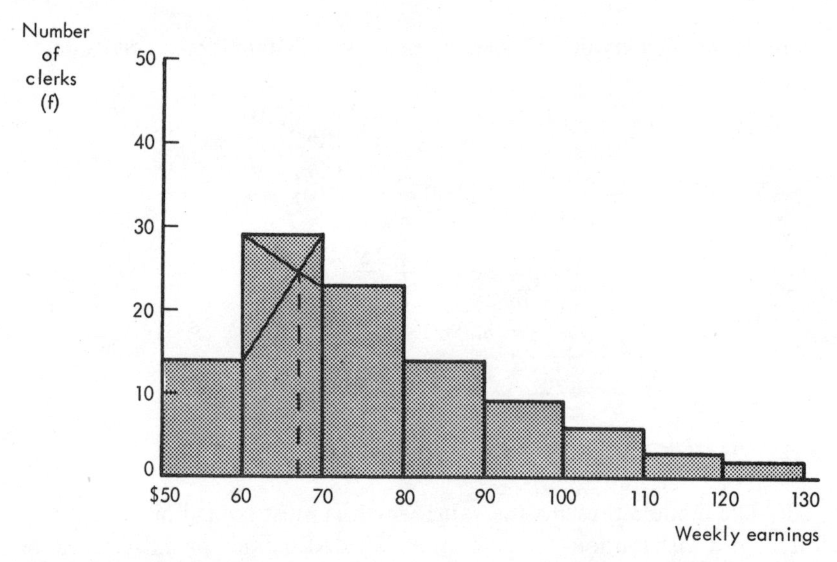

Fig. 4.2 Weekly Earnings of 100 Clerks, 1960: Histogram
SOURCE: Table 3.5.
Mode = $67.00

4.15 LOCATING THE MODE GRAPHICALLY FROM THE HISTOGRAM

The mode may be located graphically from the histogram. The procedure is as follows:

1. As indicated in Figure 4.2, the lower limit of the modal class is connected by a straight line with the lower limit of the class to the right.
2. The upper limit of the modal class is connected by a straight line with the upper limit of the class to the left.
3. From the point of intersection of the two lines, drop a perpendicular to the horizontal, or *X*, axis.
4. The approximate modal value may be read from the *X* axis. In this case, it is about $67.00.

4.16 COMPUTING THE MODE FROM THE MEAN AND THE MEDIAN

A formula for computing the mode from the mean and the median was developed by Karl Pearson (1857–1936), one of the founders of the science of statistics. He taught at the University of London; he established the statistical journal *Biometrika* and applied statistics to biological problems, especially in evolution and heredity. He developed many of the statistical techniques outlined in this text.

FORMULA:

$$M_o = \bar{x} - 3(\bar{x} - \text{Med})$$

EXAMPLE: For weekly earnings of the 100 clerks, the mode is as follows:

$$\begin{aligned} M_o &= \$76.50 - 3\,(\$76.50 - \$73.04) \\ &= \$76.50 - 3\,(\$3.46) \\ &= \$76.50 - \$10.38 \\ &= \$66.12 \end{aligned}$$

The mode obtained by this formula differs slightly from the mode computed from the previous, or difference method, formula.

4.17 RELATIONSHIP OF MEAN, MEDIAN, AND MODE IN SKEWED FREQUENCY DISTRIBUTIONS

The data for weekly earnings of clerks is a positively skewed distribution. As indicated in Figure 4.3, in a positively skewed distribution, the

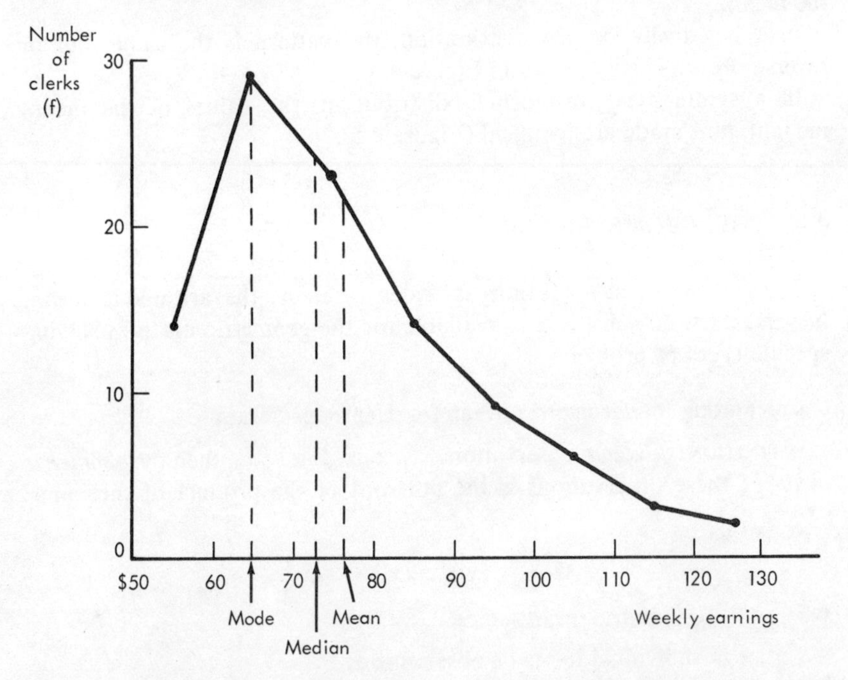

Fig. 4.3 Positively Skewed Distribution: Weekly Earnings of 100 Clerks, 1960
SOURCE: Table 3.5.

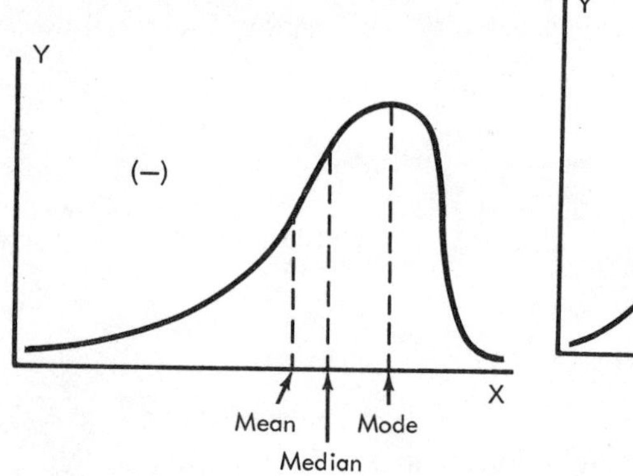

Fig. 4.4 Negatively Skewed Distribution

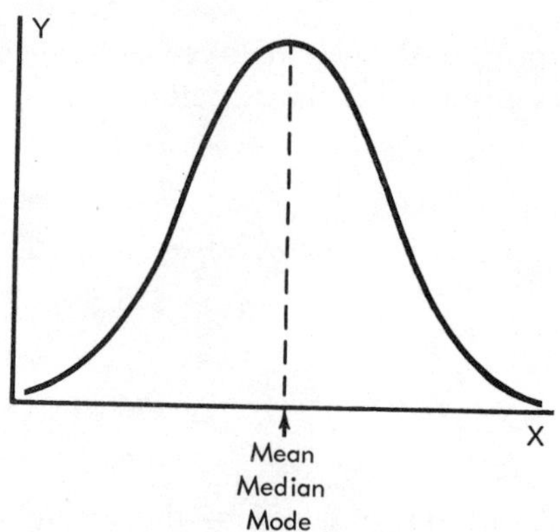

Fig. 4.5 Normal or Symmetrical Distribution

arithmetic mean is farthest toward the elongated tail of the curve, reflecting the impact of extreme values. The mode is at the maximum ordinate position, and the median lies between the mode and mean, but closer to the mean.

In a negatively skewed distribution, the pattern is the same, but in reverse order. This is shown in Figure 4.4.

In a symmetrical, or normal, distribution, the values of the mean, median, and mode are identical (Figure 4.5).

4.18 THE GEOMETRIC MEAN

The geometric mean is not as widely used as the arithmetic mean. Nevertheless, no substitute is available for the geometric mean in solving specific types of problems.

Calculating the Geometric Mean for Ungrouped Data

DEFINITION: Given n observations, $x_1, x_2, x_3, \ldots, x_n$, then the *geometric mean* of these observations is the nth root of the product of the items.

FORMULA:

$$G = \sqrt[n]{(x_1)(x_2)(x_3)\cdots(x_n)}$$

where G = geometric mean

x = individual items or observations

n = number of items in the sample

EXAMPLE 1: What is the geometric mean of 2, 5, and 9?

$$G = \sqrt[3]{(2)(5)(9)}$$
$$= \sqrt[3]{90}$$
$$= 4.48$$

This answer is computed from the logarithmic form of the formula which follows:

$$\log G = \frac{\log x_1 + \log x_2 + \log x_3 + \cdots + \log x_n}{n}$$
$$= \frac{\Sigma \log x}{n}$$

Thus,

$$\log G = \frac{\log 2 + \log 5 + \log 9}{3}$$

$$\log G = \frac{0.3010 + 0.6990 + 0.9542}{3}$$

$$\log G = \frac{1.9542}{3}$$

$$\log G = 0.6514$$

$$G = 4.48$$

EXAMPLE 2: Given the population of Queens County, New York for 1950, 1,550,849, and for 1960, 1,809,578, what is the percentage rate of population increase per year?

Solution:

$$\frac{1{,}809{,}578}{1{,}550{,}849} = 1.16683, \text{ which is the 10-year ratio of increase}^{1}$$

[1] The log and antilog in this example were derived by interpolation as follows:

$$\log 1170 = .0682 \qquad \frac{7}{10} = \frac{x}{37}$$
$$\log 1167 = x \qquad 10x = 259$$
$$\log 1160 = .0645 \qquad x = 26$$
$$.0645 + .0026 = .0671$$

$$1020 = \text{antilog } .0086 \qquad \frac{x}{10} = \frac{24}{43}$$
$$x = \text{antilog } .0067 \qquad 43x = 240$$
$$1010 = \text{antilog } .0043 \qquad x = 5$$
$$1010 + 5 = 1015$$

Obtain the annual percentage rate of increase, or x:

$$x = \sqrt[10]{1.16683}$$

Solve by logarithms as follows:

$$\log x = \frac{\log 1.16683}{10}$$

$$\log x = \frac{0.0671}{10} = 0.00671$$

The antilog of 0.00671 is 1.015. Thus $x = 1.015$. From 1.015, deduct 1.000. (The reason for this is indicated in the section, Understanding the Solution, which follows below.) Therefore, $1.015 - 1.000 = 0.015$, which, in percentage terms, is $0.015 \times 100 = 1.5\%$, the annual rate of increase per year.

UNDERSTANDING THE SOLUTION: The formula used in the solution is an adaptation of the compound interest formula as solved below:

$$l = a(1 + r)^n$$

where l = last term of the series
a = first term of the series
r = annual rate of increase
n = number of years

Therefore,

$$1{,}809{,}578 = 1{,}550{,}849(1 + r)^{10}$$

Solve by logarithms as follows:

$$\frac{\log 1{,}809{,}578 - \log 1{,}550{,}849}{10} = \log(1 + r)$$

$$\frac{6.2577 - 6.1903}{10} = \log(1 + r)$$

$$\frac{0.0674}{10} = \log(1 + r)$$

$$0.00674 = \log(1 + r)$$

The antilog of $0.00674 = 1.015 = 1 + r$

$$1.015 - 1.000 = r$$

which explains the deduction indicated in the solution.[1]
Therefore, $r = 0.015 \times 100 = 1.5\%$, the annual rate of increase per year.

[1] The antilog of 0.00674 is derived by interpolation. The computation is given in the note above.

Calculating the Geometric Mean for Grouped Data

DEFINITION: The *geometric mean* computed from grouped data is the antilog of the sum of the frequencies times the log of the class midpoints divided by n, the number of items.

FORMULA:

$$\log G = \frac{\Sigma f \log x}{n}$$

METHOD:

1. Set up the frequency distribution, including the lower and upper limits, the frequency column, and the class midpoints.
2. Set up a column for $\log x$, that is, the logarithm of each class midpoint.
3. Set up the column $f \log x$ in which the log of each class midpoint is multiplied by its corresponding frequency.
4. Compute the sum of the column, or $\Sigma f \log x$.
5. Divide $\Sigma f \log x$ by n, the number of items.
6. Compute the antilogarithm, which is the geometric mean.

4.19 CHARACTERISTICS OF THE GEOMETRIC MEAN

1. While it takes into account all the values in a set of observations, it is not as significantly affected by extreme values as the arithmetic mean.
2. The geometric mean is always smaller than the arithmetic mean.
3. When the logic of a problem suggests that a series is growing at a geometric or percentage rate, the geometric mean would yield a more meaningful measure of location than the arithmetic mean.
4. The geometric mean cannot be used as a measure of location if one of the items is zero, or if one of the values is negative.

4.20 ADVANTAGES OF THE GEOMETRIC MEAN

1. The geometric mean can be used to average ratios or percentages, while the arithmetic mean averages absolute values.
2. It is less affected by extreme values, and therefore can be used in situations where the arithmetic mean cannot.
3. It may be manipulated algebraically.
4. It is used in the computation of the daily index of spot market prices.

4.21 DISADVANTAGES OF THE GEOMETRIC MEAN

1. The geometric mean is difficult to understand, and therefore is not widely used.
2. It cannot be computed if the series contains zero or negative values.
3. The computations are more complex than those for the arithmetic mean.

4.22 THE QUARTILES

The Concept

Quartiles divide a distribution into four equal parts, with 1/4, or 25 per cent of the items in each part. This may be illustrated by Figure 4.6 where Q_1 = first quartile, Q_2 = second quartile or median, Q_3 = third quartile.

DEFINITIONS:

The *first quartile* (Q_1) is a value so determined that 1/4 of the items fall below it, and 3/4 of the items exceed it.

The *second quartile* (Q_2) is the median value.

The *third quartile* (Q_3) is a value so determined that 3/4 of the items fall below it, and 1/4 of the items exceed it.

Calculating the Quartiles for Ungrouped Data

EXAMPLE: Compute the first and third quartiles for weekly earnings of 100 clerks.

SOLUTION: The first step in computing quartiles from ungrouped data is to make an array. The data for the 100 clerks are arrayed in Table 3.2.

First quartile $(Q_1) = n/4 = 100/4 =$ *twenty-fifth entry, which equals* \$63.83.

Third quartile $(Q_3) = 3n/4 = 300/4 =$ *seventy-fifth entry, which equals* \$86.00.

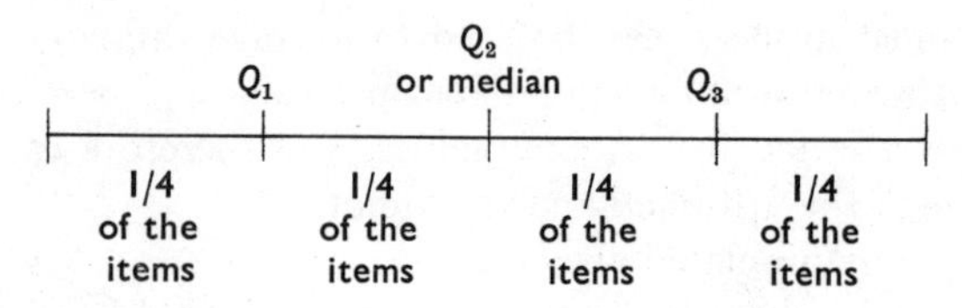

Fig. 4.6 Concept of the Quartiles

Calculating the Quartiles for Grouped Data

COMPUTATION OF FIRST QUARTILE BY COUNTING DOWN (Table 4.7): The frequency distribution is used for the computation of the first and third quartiles (Q_1 and Q_3).

EXAMPLE: Given the frequency distribution of Table 3.5, compute the first quartile (Q_1) by counting down; check on computation of first quartile by counting up; compute the third quartile (Q_3) by counting down; check on computation of the third quartile by counting up.

TABLE 4.7 CALCULATION OF THE FIRST QUARTILE (Q_1)

Weekly earnings (l_1–l_2)	Number of clerks (f)	
\$50.00–\$59.99	14	14 cumulated frequencies counting down
60.00– 69.99	29	
70.00– 79.99	23	57 cumulated frequencies counting up
80.00– 89.99	14	
90.00– 99.99	9	
100.00–109.99	6	
110.00–119.99	3	
120.00–129.99	2	
	$n = 100$	

SOURCE: Table 3.5.

Steps for computation of Q_1:	*Computation—counting down:*
1. Find $n/4$, retaining the decimal, if any.	1. $\frac{n}{4} = \frac{100}{4} = 25$
2. Count down (or count up) in the frequency column to the first quartile class, and subtract the accumulated frequencies from $n/4$.	2. $25 - 14 = 11$
3. Divide the residual thus found by the frequency of the first quartile class, and multiply by the class interval.	3. $\frac{11}{29}(\$10) = \frac{110}{29} = \$3.79$
4. Add the result to the lower limit of the first quartile class, if counting down, or subtract it from the upper limit, if counting up.	4. $\$60.00 + \$3.79 = \$63.79$

CHECK ON COMPUTATION OF FIRST QUARTILE BY COUNTING UP:

1. $\frac{3n}{4} = \frac{3(100)}{4} = \frac{300}{4} = 75$

2. $75 - 57 = 18$

3. $\frac{18}{29}(\$10) = \frac{180}{29} = \$6.206$

4. $\$69.999 - \$6.206 = \$63.79$

FORMULA FOR FIRST QUARTILE COUNTING DOWN:

$$Q_1 = l_1 + \left(\frac{\frac{n}{4} - \Sigma f_c}{f_{Q_1}}\right)$$

where l_1 = lower limit of class containing the first quartile value
n = number of items in the distribution
Σ = sum of
f_c = cumulated frequencies in the classes above when counting down
f_{Q_1} = frequency of class containing first quartile value

FORMULA FOR FIRST QUARTILE COUNTING UP:

$$Q_1 = l_2 - \left(\frac{\frac{3n}{4} - \Sigma f_c}{f_{Q_1}}\right)$$

where l_2 = upper limit of class containing the first quartile value

COMPUTATION OF THIRD QUARTILE BY COUNTING DOWN (Table 4.8):

TABLE 4.8 CALCULATION OF THE THIRD QUARTILE (Q_3)

Weekly earnings (l_1–l_2)	Number of clerks (f)	
\$50.00–\$59.99	14	66 cumulated frequencies counting down
60.00– 69.99	29	
70.00– 79.99	23	
80.00– 89.99	14	
90.00– 99.99	9	20 cumulated frequencies counting up
100.00–109.99	6	
110.00–119.99	3	
120.00–129.99	2	
	n = 100	

SOURCE: Table 3.5.

Steps:

1. $\frac{3n}{4} = \frac{3(100)}{4} = \frac{300}{4} = 75$

2. $75 - 66 = 9$

3. $\frac{9}{14}(\$10) = \frac{90}{14} = \$6.43$

4. $\$80.00 + \$6.43 = \$86.43$

CHECK ON COMPUTATION OF THIRD QUARTILE BY COUNTING UP:

1. $\frac{n}{4} = \frac{100}{4} = 25$

2. $25 - 20 = 5$

3. $\frac{5}{14}(\$10) = \frac{50}{14} = \$3.571$

4. $\$89.999 - \$3.571 = \$86.43$

4.23 PERCENTILES

THE CONCEPT: Other measures of location may be computed for other positions in an array for ungrouped data, or in a frequency distribution for grouped data. Each percent figure yields a percentile value. For example, the median value is equivalent to the fiftieth percentile. The first quartile is the twenty-fifth percentile, and the third quartile, the seventy-fifth.

DEFINITION: *Percentiles* divide a distribution into 100 parts, each of which contains 1 per cent of the items.

EXAMPLE: The thirty-seventh percentile is a position value such that 37/100 of the items fall below it, and 63/100 of the items exceed it. The interpolation method is used to obtain the value equivalent to the thirty-seventh percentile. The steps are the same as those listed for computing the median, except that in the first step the desired proportion of n is calculated.

EXERCISES

1. Define or explain the following:

Antilog	Long method for computing mean
Arithmetic mean	Mean
Class deviation	Median
Class midpoint	Modal class
Correction factor	Mode
Cumulated frequencies	More than ogive
Cumulative frequency curve	Ogive
Extreme values	Percentiles
First quartile	Position average
Geometric mean	Quartiles
Grouped data	Second quartile
Guessed mean	Short method for computing mean
Interpolation	Third quartile
Less than ogive	Ungrouped data

2. Answer the following questions relating to the mean, median, and ogive.

   a. Using the ungrouped data in Table 3.6, page 39, compute the arithmetic mean.
   b. Using the grouped data (the frequency distribution) prepared in answer to Exercise 2c, page 39:
      (1) Compute the arithmetic mean by the long method.
      (2) Compute the arithmetic mean by the short method.
   c. Compare the results obtained in answer to questions a, b(1), and b(2) above, and explain similarities or differences.
   d. Using the array prepared in answer to Exercise 2a, page 39, compute the median.
   e. Using the frequency distribution prepared in answer to Exercise 2c, page 39:
      (1) Compute the median by counting down.
      (2) Check computation of the median by counting up.
   f. Using the same frequency distribution, construct a "less than" and a "more than" cumulative frequency table.
   g. On a sheet of arithmetic graph paper:
      (1) Plot the "less than" and "more than" ogives.
      (2) Locate the median graphically.

3. Answer the following questions relating to the mode and quartiles.
   a. Using the frequency distribution prepared in answer to Exercise 2c, page 39:
      (1) Compute the mode.
      (2) Compute the first quartile by counting down.
      (3) Check the first quartile computation by counting up.
      (4) Compute the third quartile by counting down.
      (5) Check the third quartile computation by counting up.
   b. Locate the mode graphically on the histogram drawn in answer to Exercise 2e, page 39.

4. Given the frequency distribution in Table 4.9, compute the arithmetic mean, median, mode, first quartile, and third quartile.

TABLE 4.9 A FREQUENCY DISTRIBUTION

Class interval	*f*
10.0–19.9	1
20.0–29.9	3
30.0–39.9	5
40.0–49.9	9
50.0–59.9	14
60.0–69.9	9
70.0–79.9	6
80.0–89.9	2
90.0–99.9	1
Total	50

5. Using the frequency distribution in Table 4.9, construct "less than" and "more than" ogives on a sheet of arithmetic graph paper.

5

Measures of Variation

In describing a frequency distribution, in addition to analyzing measures of location, it is also important to analyze the extent of variation or variability, or, as it is sometimes called, dispersion.

The importance of a measure of variation may be illustrated by a problem in the burning life of electric light bulbs. Two questions are involved. The first is what is the average length of burning time? The second is whether the quality of the bulbs is reasonably uniform?

More specifically, do the bulbs produced by a particular manufacturer cluster closely around the arithmetic mean, as measured by hours of burning life? This would indicate uniformity of quality. Or, are individual bulbs widely dispersed around the arithmetic mean, with some bulbs burning many hours longer than the arithmetic mean and others burning many hours less than the mean?

Graphically, the alternatives may be illustrated as in Figures 5.1 *and* 5.2, *assuming both distributions are normal, and that the arithmetic mean of the hours of burning life are equal.*

It is reasonable to assume that the curve in Figure 5.1 *represents bulb production which is more uniform in quality because of the smaller amount of dispersion. Therefore, the bulbs represented by distribution A would offer less risk to the consumer as measured by burning life.*

The measures of variation or dispersion analyzed in this chapter include the following:

1. *The range*
2. *The quartile deviation* (*semi-interquartile range*)

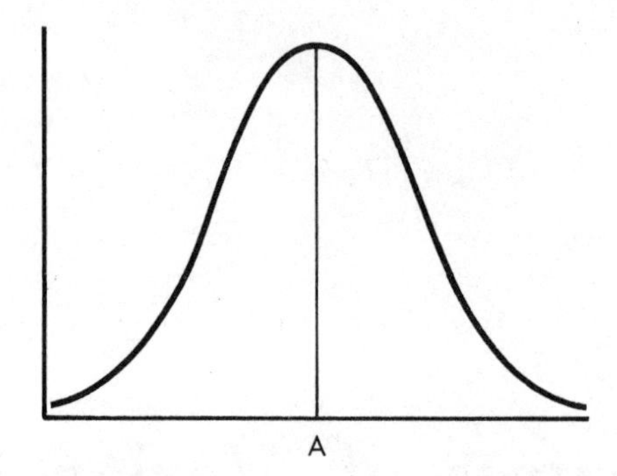

Fig. 5.1 Burning Life of Electric Light Bulb: Case A

3. *The average deviation*
4. *The standard deviation*
5. *The variance*
6. *The coefficient of variation*

The analysis for each measure includes a definition, an example with solution, the formula, and the advantages and disadvantages of each.

5.1 THE RANGE

The range is the simplest measure of dispersion. It is used to obtain a crude measure of variation both for ungrouped and grouped data.

Ungrouped Data

DEFINITION: The *range* is the difference between the lowest value and the highest value in a group of items.

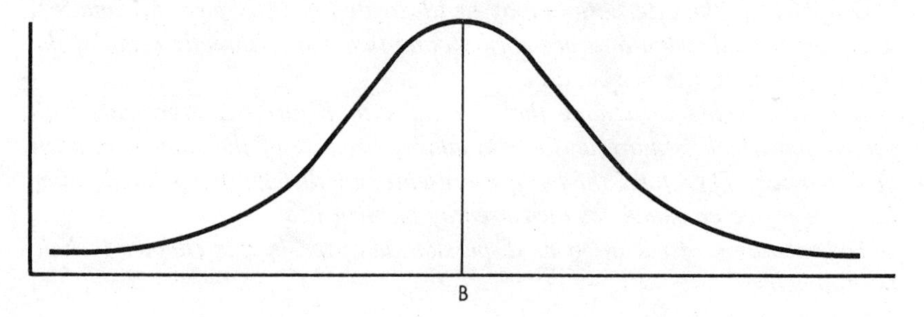

Fig. 5.2 Burning Life of Electric Light Bulb: Case B

EXAMPLE: What is the range for the weekly earnings of 100 clerks in 1960? Use the data arrayed in Table 3.2.

SOLUTION:

Note the highest value. For the 100 clerks it is . . .	$127.27
Note the lowest value. For the 100 clerks it is	51.67
The difference, or the range, is	$75.60

Grouped Data

DEFINITION: The *range* is the difference between the lower limit of the lowest value class and the upper limit of the highest value class.

EXAMPLE: What is the range for the frequency distribution of the weekly earnings of 100 clerks in 1960? Use the frequency distribution in Table 3.5.

SOLUTION:

From l_2, the upper limit of the highest value class, or . . .	$129.99
Subtract l_1, the lower limit of the lowest value class	50.00
The difference, or the range, is	$79.99

This may be rounded to $80.00.

5.2 ADVANTAGES OF THE RANGE

1. The range is the most easily computed measure of dispersion.
2. It is simple to understand.
3. It can be computed from only two values.

5.3 DISADVANTAGES OF THE RANGE

1. Because it is dependent upon the values of the extremes of a distribution, and does not take into consideration the values of the other items in the distribution, it is less useful than other measures of dispersion. In addition, this dependency upon the extreme values of a distribution may result in an unusually high value.
2. It is necessary to array ungrouped data.
3. It cannot be computed accurately for grouped data unless the lower limit of the lowest value class and the upper limit of the highest class are known.
4. It is an unstable measure of variation, because as the number of items is increased, the range also tends to increase.

5.4 THE QUARTILE DEVIATION

THE CONCEPT: The quartile deviation as a measure of variation seeks to avoid the impact of extreme values. It measures the range of the middle half, or 50 per cent of the items, the *interquartile range*, and then divides this value in half. The resulting value measures deviation from the median.

FORMULA:

$$QD = \frac{Q_3 - Q_1}{2}$$

where QD = quartile deviation, or semi-interquartile range
Q_3 = third quartile
Q_1 = first quartile

DEFINITION: The *quartile deviation* is a measure of variation which equals half the difference between the values of the first and third quartiles.

Calculating the Quartile Deviation for Ungrouped Data

EXAMPLE: Compute the quartile deviation for weekly earnings of 100 clerks. Use the data arrayed in Table 3.2.

SOLUTION: The first and third quartiles are obtained from the array in Table 3.2. The method for calculating quartiles is explained in Chapter 4.

$$Q_1 = \$63.83$$
$$Q_3 = \$86.00$$

Substituting in the formula:

$$QD = \frac{Q_3 - Q_1}{2}$$
$$= \frac{\$86.00 - \$63.83}{2}$$
$$= \frac{\$22.17}{2}$$
$$= \$11.08$$

INTERPRETATION: $\bar{x} \pm QD$ give two figures between which approximately 50 per cent of the items in a skewed distribution are included. Exactly 50 per cent of the items are included between these two values in a normal distribution.

For example, $\bar{x}$ for the ungrouped data in Table 3.2 is \$76.36. If $QD =$ \$11.08, then

$$\bar{x} + QD = \$76.36 + \$11.08 = \$87.44$$

$$\bar{x} - QD = \$76.36 - \$11.08 = \$65.28$$

Between \$65.28 and \$87.44 in the array (Table 3.2) 46 values are located, or 46 per cent of the 100 items. Between Q_3 and Q_1, known as the *inter-quartile range*, in which $Q_3 =$ \$86.00 and $Q_1 =$ \$63.83, exactly 50 items are located, or 50 per cent of the total. This is so by definition.

Although QD is relatively easy to understand, it is not widely applied in practical work.

Calculating the Quartile Deviation for Grouped Data

EXAMPLE: Compute the quartile deviation, given $Q_1 =$ \$63.79 and $Q_3 =$ \$86.43. The method for calculating quartiles from grouped data is explained in Chapter 4.

SOLUTION:

Substituting in the formula:

$$QD = \frac{Q_3 - Q_1}{2}$$

$$= \frac{\$86.43 - \$63.79}{2}$$

$$= \frac{\$22.64}{2}$$

$$= \$11.32$$

5.5 ADVANTAGES OF THE QUARTILE DEVIATION

1. The quartile deviation is easy to compute and simple to understand.
2. It can be computed from only two values, the lower value and the upper value of the middle half of the items.
3. It is more useful than the range as a rough measure of dispersion.
4. It can be calculated from an open-end distribution, or from a markedly skewed distribution.

5.6 DISADVANTAGES OF THE QUARTILE DEVIATION

1. Data gaps around the quartiles would reduce the usefulness of the quartile deviation.
2. It is too dependent upon the values of the first and third quartiles.

5.7 THE AVERAGE DEVIATION

THE CONCEPT: To obtain some idea of variation around a measure of location, it is possible to develop an average distance. This is done by subtracting each value for ungrouped data, or the midpoint of the class for grouped data, from the arithmetic mean or the median, ignoring signs, and then computing the average deviation. The reason for ignoring signs is the fact that the sum of the deviations around the arithmetic mean equals zero.

The average deviation may be computed from either the arithmetic mean or the median. The average deviation computed from the median is always less than the average deviation computed from the arithmetic mean, unless the arithmetic mean equals the median.

Calculating the Average Deviation for Ungrouped Data

DEFINITION: The *average deviation* is a measure of variation from the arithmetic mean or the median which equals the sum of the deviations, signs ignored, of the items from the arithmetic mean, or median, divided by the number of items.

FORMULA:

$$AD = \frac{\Sigma\,|x - \bar{x}|}{n}$$

where AD = average deviation

Σ = sum of

$|x - \bar{x}|$ = deviation of each x item from arithmetic mean of the items, without regard to sign

$|\ \ |$ = the short, vertical, parallel lines indicate without regard to sign

n = number of items in sample

EXAMPLE: Given final examination grades, 90, 90, 85, 80, and 75, compute the average deviation.

SOLUTION: First, compute the arithmetic mean:

$$\bar{x} = \frac{\Sigma x}{n}$$

$$= \frac{420}{5}$$

$$= 84$$

Second, set up a table similar to Table 5.1:

TABLE 5.1 FIVE EXAMINATION GRADES: CALCULATION OF THE AVERAGE DEVIATION

x	$\lvert x - \bar{x} \rvert$
90	6
90	6
85	1
80	4
75	9
420	$\Sigma \lvert x - \bar{x} \rvert = 26$

To obtain items for column $|x - \bar{x}|$, the arithmetic mean is deducted from each x item. For example, $90 - 84 = 6$. Signs are ignored.

Third, column $|x - \bar{x}|$ is added to obtain $\Sigma |x - \bar{x}|$.

Fourth, the average deviation is computed as follows:

$$AD = \frac{\Sigma |x - \bar{x}|}{n}$$

$$= \frac{26}{5}$$

$$= 5.2$$

Calculating the Average Deviation for Grouped Data

DEFINITION: The *average deviation* is a measure of variation from the arithmetic mean or the median; it equals the sum of the frequencies times the deviation of the class midpoint from the arithmetic mean, or the median, of the series, signs ignored, divided by the number of items.

FORMULA:

$$AD = \frac{\Sigma f\,|x - \bar{x}|}{n}$$

where AD = average deviation
Σ = sum of
f = frequency
$|x - \bar{x}|$ = deviation of each x item from arithmetic mean of the items without regard to sign
$|\ \ |$ = short, vertical, parallel lines indicate without regard to sign
n = number of items in sample

EXAMPLE: Given the frequency distribution of the weekly earnings of 100 clerks in 1960, compute the average deviation.

SOLUTION: Using the data in Table 3.5, set up Table 5.2.

TABLE 5.2 CALCULATION OF THE AVERAGE DEVIATION

Weekly earnings (l_1–l_2) (1)	Number of clerks (f) (2)	Class midpoint x (3)	$\|x - \bar{x}\|$ (4)	$f\|x - \bar{x}\|$ (5)
$50.00–$59.99	14	$ 55	$21.50	$ 301.00
60.00– 69.99	29	65	11.50	333.50
70.00– 79.99	23	75	1.50	34.50
80.00– 89.99	14	85	8.50	119.00
90.00– 99.99	9	95	18.50	166.50
100.00–109.99	6	105	28.50	171.00
110.00–119.99	3	115	38.50	115.50
120.00–129.99	2	125	48.50	97.00
	n = 100			$1,338.00

SOURCE: Table 3.5.

First, compute the arithmetic mean: $\bar{x}$ = $76.50.

Second, to obtain values for column 4, each x value, or class midpoint, is deducted from the arithmetic mean.

Third, to obtain values for column 5, multiply column 2 values by column 4 values.

Fourth, obtain the sum of column 5, and then divide by n as follows:

$$AD = \frac{\Sigma f\,|x - \bar{x}|}{n}$$

$$= \frac{\$1{,}338.00}{100}$$

$$= \$13.38$$

5.8 ADVANTAGES OF THE AVERAGE DEVIATION

1. The average deviation gives equal weight to the deviation of every value from the mean or the median.
2. The average deviation may be computed from either the arithmetic mean or the median.
3. It has more applicability than the range or quartile deviation.
4. It is less affected by extreme values than the standard deviation.

5.9 DISADVANTAGES OF THE AVERAGE DEVIATION

1. The average deviation ignores minus signs, and therefore cannot be manipulated algebraically.
2. It is not adapted for further statistical analysis.

5.10 THE STANDARD DEVIATION

THE CONCEPT: Standard deviation is the measure of variation most widely used in statistical analysis. It measures how far the weighted relationship of each of the items in a frequency distribution is located from the arithmetic mean. If a frequency distribution is spread out, or widely dispersed, the standard deviation is large. If a frequency distribution is closely packed, or narrowly dispersed, the standard deviation is small. When the standard deviation is small, the arithmetic mean is a meaningful measure of central tendency.

POPULATION VS SAMPLE: At this point it is well to distinguish between a population, or universe as it is sometimes called, and a sample.

STANDARD DEVIATION SYMBOLS: The small Greek letter *sigma*, σ, is used to designate the standard deviation applicable to a statistical *population*, while the letter, s, is used to designate the standard deviation of a *sample*.

DEFINITIONS:

A *population* is the totality of all possible observations, objects, individuals, or things under consideration.

A *sample* is a segment or part of a population selected in a particular way. A sample is often referred to in statistics as the items or values which have been observed or selected from a population. Statisticians are generally concerned with sample data.

For example, in 1960 about 53 million households were counted in the United States Census. This total may be thought of as a *population*. If 25,000 of these households were selected for study, these 25,000 households would constitute the *sample*.

Calculating the Standard Deviation for Ungrouped Data

DEFINITION: The *standard deviation* is a measure of variation from the arithmetic mean. It equals the square root of the average of the squared deviations of the items from the arithmetic mean.

FORMULA:

$$s = \sqrt{\frac{\Sigma (x - \bar{x})^2}{n}}$$

where s = standard deviation

Σ = sum of

x = individual item

$\bar{x}$ = arithmetic mean

n = number of items in sample

EXAMPLE: Given five examination grades, 90, 90, 85, 80, and 75, compute the standard deviation (Table 5.3).

TABLE 5.3 FIVE EXAMINATION GRADES: CALCULATION OF THE STANDARD DEVIATION

x	$(x - \bar{x})$	$(x - \bar{x})^2$
90	+6	36
90	+6	36
85	+1	1
80	−4	16
75	−9	81
420	0	170

First, the arithmetic mean is computed.

$$\bar{x} = \frac{\Sigma x}{n}$$

$$= \frac{420}{5}$$

$$= 84$$

Second, to obtain items for column $(x - \bar{x})$, the arithmetic mean is deducted from each x item. For example, $90 - 84 = +6$.

Third, the deviation of each x item from the mean is then squared to obtain the third column in the table, $(x - \bar{x})^2$.

Fourth, the values derived in the table are substituted in the formula as follows:

$$s = \sqrt{\frac{\Sigma (x - \bar{x})^2}{n - 1}}$$

$$= \sqrt{\frac{170}{4}}$$

$$= \sqrt{42.5}$$

$$= 6.519$$

When the number of items in the sample is less than 30, $n - 1$ is used in the denominator instead of n. This is explained in Chapter 7 in the discussion of degrees of freedom.

Calculating the Standard Deviation for Grouped Data

FORMULA:

$$s = i\sqrt{\frac{\Sigma f(d)^2}{n} - \left(\frac{\Sigma fd}{n}\right)^2}$$

where s = standard deviation

Σ = sum of

f = frequency

d = class deviation

n = number of items in sample

EXAMPLE: Given the frequency distribution of weekly earnings of 100 clerks, as indicated in Table 3.5, compute the standard deviation.

SOLUTION: Set up Table 5.4.

TABLE 5.4 CALCULATION OF THE STANDARD DEVIATION

Weekly earnings (l_1–l_2)	Number of clerks (f)	Class midpoint (x)	d	fd		$f(d)^2$
\$50.00–\$59.99	14	\$ 55	−2	−28	−57	56
60.00– 69.99	29	65	−1	−29		29
70.00– 79.99	23	75	0	0		0
80.00– 89.99	14	85	+1	+14	+72	14
90.00– 99.99	9	95	+2	+18		36
100.00–109.99	6	105	+3	+18		54
110.00–119.99	3	115	+4	+12		48
120.00–129.99	2	125	+5	+10		50
	$n = 100$			$\Sigma fd = +15$		$287 = \Sigma f(d)^2$

SOURCE: Table 3.5.

Substitute values in the formula.

$$s = i\sqrt{\frac{\Sigma f(d)^2}{n} - \left(\frac{\Sigma fd}{n}\right)^2}$$

$$= \$10\sqrt{\frac{287}{100} - \left(\frac{15}{100}\right)^2}$$

$$= \$10\sqrt{2.87 - \frac{225}{10{,}000}}$$

$$= \$10\sqrt{2.87 - .0225}$$

$$= \$10\sqrt{2.8475}$$

$$= (\$10)(1.687)$$

$$= \$16.87$$

5.11 THE RELATIONSHIP BETWEEN $\bar{x}$ AND s

The standard deviation, s, provides limits of variation around the arithmetic mean, $\bar{x}$, This relationship is expressed as follows:

$$\bar{x} \pm 1s$$

$$\bar{x} \pm 2s$$

$$\bar{x} \pm 3s$$

EXAMPLE: Weekly earnings for the 100 clerks in 1960 have been analyzed, yielding the following results:

$$\bar{x} = \$76.50, \quad \text{and} \quad s = \$16.87.$$

The limits for $\bar{x} \pm 1s$ are $76.50 + $16.87, or $93.37, and $76.50 − $16.87, or $59.63. Thus, $\bar{x} \pm 1s = \$59.63$ to $93.37.

$\pm 2s = (2)(\$16.87)$, or $33.74

$$\bar{x} + 2s = \$76.50 + \$33.74 = \$110.24$$

$$\bar{x} - 2s = \$76.50 - \$33.74 = \$42.76$$

$$\bar{x} \pm 2s = \$42.76 \text{ to } \$110.24$$

$\pm 3s = (3)(\$16.87)$, or $50.61

$$\bar{x} + 3s = \$76.50 + \$50.61 = \$127.11$$

$$\bar{x} - 3s = \$76.50 - \$50.61 = \$25.89$$

$$\bar{x} \pm 3s = \$25.89 \text{ to } \$127.11$$

5.12 INTERPRETATION OF THE STANDARD DEVIATION

The standard deviation of a sample of observations is a very valuable tool for the statistician. Using the data above, assuming the universe from which the sample was selected is normally distributed, and assuming that the sample is representative of the population, then the following relationships would be approximately applicable: The range of the mean plus and minus one standard deviation would contain 68.27 per cent of the observations; the mean plus and minus two standard deviations would contain 95.45 per cent of the observations; and the mean plus and minus three standard deviations would contain 99.73 per cent of the observations.

5.13 ADVANTAGES OF THE STANDARD DEVIATION

1. The standard deviation is the easiest measure of disperison to handle algebraically, and is the one most widely used.
2. It is very useful for further statistical analysis.
3. It can be computed approximately by taking one-sixth of the range.

5.14 DISADVANTAGES OF THE STANDARD DEVIATION

1. The standard deviation is more difficult to compute than the other measures of dispersion.
2. It is more difficult to understand.
3. It is significantly affected by extreme values.

5.15 THE VARIANCE

THE CONCEPT: The variance is the square of the standard deviation.

DEFINITION: The *variance* is the arithmetic mean of the squared differences from the mean of a distribution.

FORMULA FOR A POPULATION:

$$\sigma^2 = \frac{\Sigma (x - \mu)^2}{N}$$

where σ^2 = variance
Σ = sum of
x = individual item
μ = arithmetic mean of population
N = number of items in population

FORMULA FOR A SAMPLE:

$$s^2 = \frac{\Sigma (x - \bar{x})^2}{n}$$

where s^2 = variance
Σ = sum of
x = individual item
$\bar{x}$ = arithmetic mean of sample
n = number of items in sample

5.16 CHARACTERISTICS OF THE VARIANCE

1. Its value is small when the items in a distribution cluster closely around the arithmetic mean, and large when the items are widely dispersed.
2. It is a value which is not affected by the number of items in the distribution.

3. Adding a constant to each variable does not change the variance.

4. Multiplying each variable by a constant is the same as multiplying the variance by the square of that constant.

5. If two variables are independent, then the variance of the two variables is equal to the sum of the variance of each variable.

6. The variance of two or more combined distributions can be calculated from the variance and the means of the respective distributions.

7. The characteristics of the variance are shared also by the standard deviation, since the standard deviation is simply the square root of the variance.

5.17 APPLICATION OF THE STANDARD DEVIATION AND THE VARIANCE

The standard deviation and the variance are measures of variation or dispersion. For any series, the standard deviation and the variance measure variability among the items. The greater the variability, the greater is the standard deviation and the variance, and vice versa.

5.18 THE COEFFICIENT OF VARIATION

THE CONCEPT: The standard deviation is a measure of absolute variation in a frequency distribution. However, it is affected by the scale of the variable, such as inches or pounds. Therefore, when a comparison is to be made between variables of different scales, dividing the standard deviation by the mean of the variable cancels out the scale effect and provides a meaningful comparison of variation or dispersion between two series. The value derived is known as a measure of relative variation or dispersion.

For example, do fifteen-year-old boys vary more in weight than in height? Pounds are to be compared with inches, and a measure of relative dispersion or variation is required.

The most commonly used measure of relative variation is called the *coefficient of variation.*

DEFINITION: The *coefficient of variation* is a measure of relative dispersion which translates the standard deviation into a percentage of the arithmetic mean.

FORMULA:

$$V = \frac{s}{\bar{x}}(100)$$

where V = coefficient of variation

s = standard deviation

$\bar{x}$ = arithmetic mean

EXAMPLE: Calculate the coefficient of variation for weekly earnings of 100 clerks, using the data of Table 3.5.

SOLUTION: Substitute values computed previously in the formula for the coefficient of variation.

$$V = \frac{s}{\bar{x}}(100)$$

$$= \frac{\$16.87}{\$76.50}(100)$$

$$= 22.0\%$$

Usually, the numerical value for the coefficient of variation is compared with the coefficient of variation of one or more other series. It has no direct application by itself.

5.19 ADVANTAGES OF THE COEFFICIENT OF VARIATION

1. If two series of data are expressed in the same or in different units and a comparison is desired, the coefficient of variation provides the necessary tool.
2. It is expressed as a pure number that is independent of the unit of measurement.

EXERCISES

1. Define or explain the following:

Average deviation	Range
Coefficient of variation	Relative variation
Deviation	Sample
Dispersion	Semi-interquartile range
Interquartile range	Standard deviation
Population	Variance
Quartile deviation	Variation

2. Answer the following:
   a. Using the frequency distribution prepared in answer to Exercise 2c, page 39, compute:
      (1) The range.
      (2) The quartile deviation (the semi-interquartile range).
      (3) The standard deviation
      (4) The variance
      (5) The coefficient of variation.
   b. Compute the coefficient of variation for a sample of controlled rents which have an arithmetic mean of $59.07 and a standard deviation of $24.50.
   c. Using the coefficients of variation computed in answer to Exercise 2a(5) and 2b above, explain which coefficient indicates greater variation.
3. Using the frequency distribution of Table 4.9, page 65, compute the standard deviation.
4. Using the frequency distribution of Table 4.9, page 65, compute the quartile deviation and the average deviation.

6

Probability

In this chapter the student is introduced to techniques by which he can generalize from sample data that have been summarized by methods previously discussed. The student now enters the realm of decision-making in which a choice must be made between alternatives which are characterized by uncertainty. Thus, the concept of statistics is expanded to encompass a wider range of problems. Examples of such problems occur in everyday living. The typical person must choose between alternatives each of which is subject to uncertainty. Should he take an umberella or not? Which one of several alternative routes of travel will bring him to his office in the shortest possible time on a particular morning?

But uncertainty is inherent in a multitude of other situations. For example, the essence of games of chance is uncertainty; or the outcome of an election; or experiments with spacecraft; or use of various drugs in treating disease; or government credit policies; or timing in the purchase of raw materials.

Modern statistics, as it is known today, started as a tool of gamblers who sought more precise estimates of odds at the gaming tables. Galileo (1564–1642) *and Blaise Pascal* (1623–1662) *made significant contributions. Thus, the prodding of gamblers concerned with dice, cards, and the roulette wheel helped to stimulate thinking about the theory of probability, the rules of probability, and the laws of chance.*

In the subsequent three centuries, the science of statistics grew, benefiting from the contributions of such scholars as Abraham De Moivre (1667–1754), *Marquis de Laplace* (1749–1827), *Carl Friedrich Gauss* (1777–1855), *and Francis Galton* (1822–1911). *In more recent times, Karl Pearson*

(1857–1936), *William S. Gosset* (1876–1937), *and R. A. Fisher* (1890–1962) *made notable contributions. The development of econometrics in this and the previous century rests heavily on the work of Augustin Cournot* (1801–1877), *Leon Walras* (1834–1910), *Vilfredo Pareto* (1848–1923), *and Alfred Marshall* (1842–1924).

This chapter analyzes the concept of probability, the types of probability, and the rules of probability, as illustrated by an experiment in dice tossing. After reviewing the concepts of factorials, permutations, combinations, and the binomial expansion, the chapter analyzes the binomial probability distribution, and then the normal probability distribution as the limit of the binomial.

The standard form of the normal curve follows along with the relationship between the arithmetic mean and standard deviation in a population. The z transformation then illustrates the technique for changing from the X scale to the z scale, and vice versa. Several examples illustrate the technique.

The chapter concludes with a brief explanation of other discrete and continuous probability distributions.

6.1 THE CONCEPT OF PROBABILITY

Probability may be thought of as a quantitative value which is generally expressed as a ratio. The probability of an event which cannot happen under any circumstances is zero. If the probability of the occurrence of an event is certain, then its probability of occurrence is one.

The use of a ratio to represent a probability is not peculiar to statisticians. In sports, the word odds is used. For example, a fan may predict that a team has three chances to two of vanquishing the opponents. The expression 3:2 is referred to as *odds*. These are converted to probabilities by forming fractions with these numbers as numerators and their sum as the common denominator.

For example, odds of 3:2, or the probability of the local team winning, are:

$$\frac{3}{3+2} = \frac{3}{5} = 0.6$$

The probability of the opposing team winning is

$$\frac{2}{3+2} = \frac{2}{5} = 0.4$$

The sum of both probabilities (winning and losing) is $0.6 + 0.4 = 1$, assuming that a tie game is not permitted.

Thus, conceptually, probability may be viewed as follows:

$$\textit{Probability} = \frac{\textit{number of successes}}{\textit{total number of events (or successes + failures)}}$$

DEFINITION: *Probability* is the proportion of times that a given event can be expected to occur in an infinite number of attempts. If an event can occur in n equally likely ways and n_a ways have the property a, then the probability that a will occur equals na/n.

6.2 TYPES OF PROBABILITY

The types of probability are *a priori* probability and *empirical* probability.

1. An *a priori* probability may be computed prior to trial or experimentation. For example, the probability of obtaining heads in tossing a coin is 1/2.
2. An *empirical* probability must be estimated on the basis of observation and experimentation.

6.3 RULES OF PROBABILITY

The computation of probability follows certain basic rules. These rules are easier to explain and to understand when applied to games of chance. They will be illustrated, therefore, in a dice-tossing experiment. The two basic rules are the *multiplication rule of independent events* and the *addition rule.*

1. The *multiplication rule of independent events:* If two events are *independent*, the probability of their joint occurrence is the product of their individual probabilities.

Two or more events are *independent* when the occurrence or nonoccurrence of one event does not affect the occurrence of any of the other events. For example, when two dice are tossed, getting an ace (that is, a one) on the first die is independent of getting an ace on the second die.

2. The *addition rule:* The probability of the occurrence of one or the other of two *mutually exclusive events* is the sum of their individual probabilities.

Two or more events are *mutually exclusive* if the occurrence of one precludes the occurrence of the others. For example, in tossing a single die, any numbered face from 1 to 6 may turn up. Only one number will turn up, and all the rest are automatically precluded from occurring, assuming a single toss.

TABLE 6.1 PROBABILITIES FOR OBTAINING SPECIFIED POINTS IN TOSSING TWO DICE

	Possible combinations		Probability of			
			Obtaining each specific die face for		Obtaining each combination (multiplication rule)	Obtaining specified point (addition rule)
Point required	Die A	Die B	Die A	Die B		
2	1	1	1/6	1/6	(1/6)(1/6) = 1/36	1/36
3	1	2	1/6	1/6	(1/6)(1/6) = 1/36	
	2	1	1/6	1/6	(1/6)(1/6) = 1/36	1/36 + 1/36 = 2/36
4	1	3	1/6	1/6	(1/6)(1/6) = 1/36	
	2	2	1/6	1/6	(1/6)(1/6) = 1/36	
	3	1	1/6	1/6	(1/6)(1/6) = 1/36	1/36 + 1/36 + 1/36 = 3/36
5	1	4	1/6	1/6	(1/6)(1/6) = 1/36	
	2	3	1/6	1/6	(1/6)(1/6) = 1/36	
	3	2	1/6	1/6	(1/6)(1/6) = 1/36	1/36 + 1/36
	4	1	1/6	1/6	(1/6)(1/6) = 1/36	+ 1/36 + 1/36 = 4/36
6	1	5	1/6	1/6	(1/6)(1/6) = 1/36	
	2	4	1/6	1/6	(1/6)(1/6) = 1/36	
	3	3	1/6	1/6	(1/6)(1/6) = 1/36	
	4	2	1/6	1/6	(1/6)(1/6) = 1/36	1/36 + 1/36 + 1/36
	5	1	1/6	1/6	(1/6)(1/6) = 1/36	+ 1/36 + 1/36 = 5/36
7	1	6	1/6	1/6	(1/6)(1/6) = 1/36	
	2	5	1/6	1/6	(1/6)(1/6) = 1/36	
	3	4	1/6	1/6	(1/6)(1/6) = 1/36	
	4	3	1/6	1/6	(1/6)(1/6) = 1/36	1/36 + 1/36
	5	2	1/6	1/6	(1/6)(1/6) = 1/36	+ 1/36 + 1/36
	6	1	1/6	1/6	(1/6)(1/6) = 1/36	+ 1/36 + 1/36 = 6/36
8	2	6	1/6	1/6	(1/6)(1/6) = 1/36	
	3	5	1/6	1/6	(1/6)(1/6) = 1/36	
	4	4	1/6	1/6	(1/6)(1/6) = 1/36	
	5	3	1/6	1/6	(1/6)(1/6) = 1/36	1/36 + 1/36 + 1/36
	6	2	1/6	1/6	(1/6)(1/6) = 1/36	+ 1/36 + 1/36 = 5/36
9	3	6	1/6	1/6	(1/6)(1/6) = 1/36	
	4	5	1/6	1/6	(1/6)(1/6) = 1/36	
	5	4	1/6	1/6	(1/6)(1/6) = 1/36	1/36 + 1/36
	6	3	1/6	1/6	(1/6)(1/6) = 1/36	+ 1/36 + 1/36 = 4/36
10	4	6	1/6	1/6	(1/6)(1/6) = 1/36	
	5	5	1/6	1/6	(1/6)(1/6) = 1/36	
	6	4	1/6	1/6	(1/6)(1/6) = 1/36	1/36 + 1/36 + 1/36 = 3/36
11	5	6	1/6	1/6	(1/6)(1/6) = 1/36	
	6	5	1/6	1/6	(1/6)(1/6) = 1/36	1/36 + 1/36 = 2/36
12	6	6	1/6	1/6	(1/6)(1/6) = 1/36	1/36

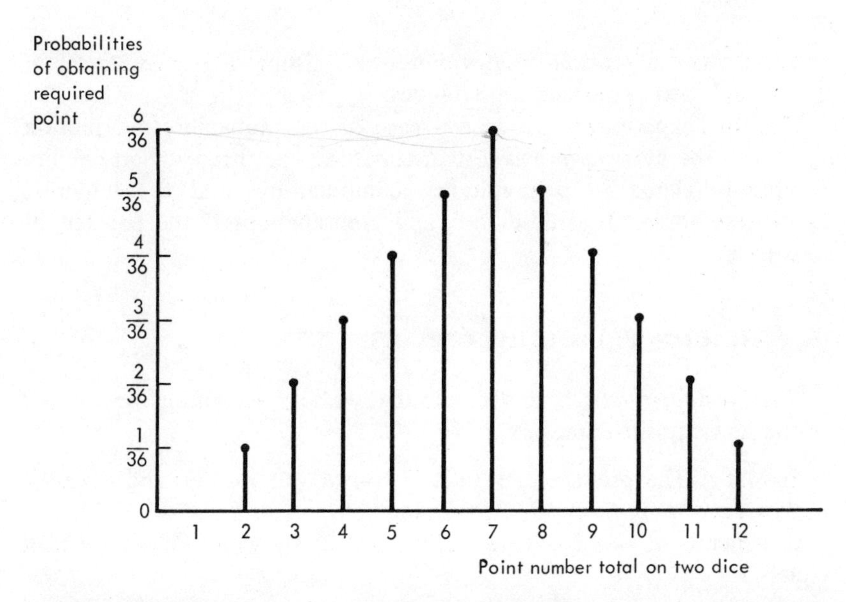

Fig. 6.1 Distribution of Probabilities of Obtaining Specified Points When Rolling Two Dice

EXAMPLE: In a toss of two dice, what is the probability of obtaining each of the following: 2, 3, 4, 5, 6, 7, 8, 9, 10, 11, and 12? Tables 6.1 and 6.2 and Figure 6.1 indicate these probabilities. They present data which make up a probability distribution.

TABLE 6.2 SUMMARY OF PROBABILITIES OF OBTAINING SPECIFIED POINTS IN TOSSING TWO DICE

Point number	Number of ways of obtaining required point	Probability of obtaining specified point
2	1	1/36
3	2	2/36
4	3	3/36
5	4	4/36
6	5	5/36
7	6	6/36
8	5	5/36
9	4	4/36
10	3	3/36
11	2	2/36
12	1	1/36
Total	36	36/36 or 1

DEFINITION: A *probability distribution* is a listing of a possible set of events and their associated probabilities.

The dice experiment is a special case of the probability distribution known as the *binomial probability distribution.* The theory of probability is closely related to permutations, combinations, and the binomial expansion. Basic to permutations and combinations is the concept of factorials.

6.4 THE CONCEPT OF FACTORIALS

The word *factorial* is both a term and a symbol representing the product of the first n positive integers.

EXAMPLE: The product of (1) (2) (3) (4) (5) (6) = 720, and is called *factorial six* or *six factorial.* The symbol is 6!

In general, $n! = n(n - 1)(n - 2) \ldots (3)(2)(1)$ while $1! = 1$, $0!$ also equals 1.

6.5 PERMUTATIONS

DEFINITION: A *permutation* is an arrangement of things where the order or arrangement is important.

EXAMPLE: Given the three letters, A, B, and C, how many permutations can be made? The answer is 6, derived as follows:

ABC BAC CAB

ACB BCA CBA

Thus, there are 6 permutations of 3 things when they are all considered together, or $n!$ ways of arranging n things.

FORMULA:

$$_nP_n = n(n - 1)(n - 2) \ldots [n - (n - 1)] = n!$$

The number of permutations of n things taken n at a time equals n factorial.

Suppose it is not desired to consider all n things together, but rather to consider the permutations of n things taken r at a time, where r is equal to or less than n.

EXAMPLE: Assume that a salesman is instructed to visit any 3 of the 5 cities, A, B, C, D, and E, and in any order. How many different journeys has he to choose from?

FORMULA[1]:

$$_nP_r = n(n-1)(n-2)\ldots(n-r+1)$$

or

$$_nP_r = \frac{n!}{(n-r)!}$$

Substituting in the formula,

$$_5P_3 = \frac{5!}{2!} = \frac{(5)(4)(3)(2)(1)}{(2)(1)} = 60$$

EXAMPLE: Given three letters, A, B, and C, how many permutations can be made taking two letters at a time?

$$_nP_r = \frac{n!}{(n-r)!}$$

$$_3P_2 = \frac{3!}{1!} = \frac{(3)(2)(1)}{1} = 6$$

6.6 COMBINATIONS

DEFINITION: A *combination* is a grouping of things, the order of which is not important.

EXAMPLE 1: Consider the three letters A, B, and C. How many combinations may be made of them?

The answer is 1, because ABC is the same as BAC or CAB. The order of the letters is not important.

If it is not desired to consider all n things at the same time, the problem is solved by means of the following formula.

EXAMPLE 2: Given the three letters A, B, and C, how many combinations may be made using two letters at a time?

FORMULA[2]:

$$_nC_r = \frac{n!}{r!\,(n-r)!}$$

$$_3C_2 = \frac{3!}{2!\,1!} = \frac{(3)(2)(1)}{(2)(1)(1)} = \frac{6}{2} = 3$$

Actually, a problem as simple as this may be solved by observation. The three letter combinations which may be formed are AB, AC, and BC. A more complex problem follows.

[1] Alternative notations for the permutation symbol include P_r^n and Pn,r.

[2] Alternative notations for the combination symbol include C_r^n and Cn,r.

EXAMPLE 3: How many combinations may be made of 10 things taken 4 at a time?

$$
{}_{10}C_4 = \frac{10!}{4!\,6!} = \frac{\overset{5}{(\cancel{10})}\overset{3}{(\cancel{9})}\overset{2}{(\cancel{8})}(7)\overset{1}{(\cancel{6!})}}{\underset{1}{(\cancel{4})}\underset{1}{(\cancel{3})}\underset{1}{(\cancel{2})}(1)\underset{1}{(\cancel{6!})}} = 210
$$

6.7 THE BINOMIAL EXPANSION

CONCEPT OF THE BINOMIAL THEOREM: The binomial theorem provides a short method for *expanding* positive, integral powers of the binomial $(x + y)$.

EXAMPLE:

$$
\begin{aligned}
(x + y)^1 &= x + y \\
(x + y)^2 &= x^2 + 2xy + y^2 \\
(x + y)^3 &= x^3 + 3x^2y + 3xy^2 + y^3 \\
(x + y)^4 &= x^4 + 4x^3y + 6x^2y^2 + 4xy^3 + y^4
\end{aligned}
$$

and so forth.

The general formula for the binomial expansion is as follows:

$$
(x + y)^n = x^n + \frac{n}{1}x^{n-1}y + \frac{n(n-1)}{(1)(2)}x^{n-2}y^2 + \ldots + y^n
$$

Pascal's Triangle

The coefficients for the various binomials may be obtained from the triangle developed by Blaise Pascal (1623–1662) and shown in Figure 6.2. Each coefficient in the table may be derived by adding together the two terms of the previous line, one term from the left and one from the right. For example, when $n = 7$, the coefficient 35 is the sum of 15 and 20.

Fig. 6.2 Pascal's Triangle

Number in sample n	Binomial coefficients	Sum
1	1 1	2
2	1 2 1	4
3	1 3 3 1	8
4	1 4 6 4 1	16
5	1 5 10 10 5 1	32
6	1 6 15 20 15 6 1	64
7	1 7 21 35 35 21 7 1	128
8	1 8 28 56 70 56 28 8 1	256
9	1 9 36 84 126 126 84 36 9 1	512
10	1 10 45 120 210 252 210 120 45 10 1	1,024

Probabilities can be determined from the foregoing summary of the binomial expansion. The characteristics which must be satisfied are

1. The events can be classified into only two categories, as success-failure, defective-nondefective.
2. The probability of a success remains constant from one trial to another.
3. The events are independent.

If the events are independent, the probability of occurrence can be found by expanding the binomial $(p + q)^n$, where p is the proportion of successes, q is the complement of p, or the proportion of failures, and n is the number of items in the sample.

EXAMPLE: In the production of machine stampings, a firm finds that 60 per cent are top grade, and 40 per cent are average grade. What is the probability of selecting a top grade and an average grade stamping in a sample of two?

SOLUTION: Given $n = 2$, $p = .6$, and $q = .4$.

This solution is based upon expansion of the binomial $(p + q)^n$, or in this example, $(p + q)^2$ which equals $p^2 + 2pq + q^2$.

Substituting $p = .6$ and $q = .4$, results in

$$(.6 + .4)^2 = .6^2 + 2(.6)(.4) + .4^2$$
$$= .36 + .48 + .16$$

Thus, the probability of 2 top grade stampings = 0.36
the probability of 1 top grade stamping = 0.48
and the probability of 2 average grade stampings = 0.16
Total probability = 1.00

6.8 THE BINOMIAL PROBABILITY DISTRIBUTION

The binomial probability distribution is of central importance in statistical theory and in the application of statistical methods. It rests heavily on an understanding of permutations, combinations, and the expansion of a binomial.

DEFINITION: The *binomial distribution* is one which is used to determine probabilities when events can be classified into only two categories, such as, success-failure, defective-nondefective, male-female, yes-no; when the trials are independent; and when the probability remains constant from trial to trial.

FORMULA: The general formula of the binomial distribution is as follows:

$$y = \frac{n!}{x!\,(n-x)!}\,p^x q^{n-x}$$

where n = number of independent events in a trial
p = probability of success in a single event
q = probability of failure
x = stated number of successes
y = probability of obtaining the stated number of successes
$n!$ = factorial n, or product of the integers from 1 to n
$x!$ = factorial x

UNDERSTANDING THE FORMULA: The formula above consists of two parts:

1. The first part $n!/x!\,(n-x)!$ is the binomial coefficient. It counts the number of different arrangements of exactly x successes and $n - x$ failures in n trials.
2. The second part, the factor $p^x q^{n-x}$, is the probability for a specific arrangement of x successes and $n - x$ failures.

Binomial probabilities are always products of these two parts. For example, if $x = 3$ and $n - x = 2$, then $p^x q^{n-x}$ is the probability of getting a specific arrangement of three successes and two failures, while $n!/x!\,(n-x)!$ is the number of ways of obtaining exactly three successes and two failures in five trials.

EXAMPLE 1: What is the probability of obtaining three aces (ones) in rolling a die three times? In this problem,

$$y = 3$$
$$p = 1/6$$
$$q = 5/6$$
$$x = 3$$
$$n - x = 0$$
$$y = \text{probability of obtaining 3 aces}$$

Substituting in the formula,

$$y = \frac{3!}{3!\,0!}\,(1/6)^3\,(5/6)^0$$

Since $0! = 1$, and $(5/6)^0 = 1$, then

$$y = (1/6)^3 \text{ or } (1/6)\,(1/6)\,(1/6) \text{ or } 1/216$$

Thus, the probability of obtaining three aces in rolling a die three times is 1/216.

EXAMPLE 2: A bowl contains 4 red balls and 6 green balls. What is the probability of obtaining exactly 2 red and 3 green balls in drawing 5 balls from the bowl? Assume that each ball is replaced in the bowl as soon as it is selected and identified as to color.

The probability of drawing a red ball is $4/10 = 2/5 = .4$.
The probability of drawing a green ball is $6/10 = 3/5 = .6$.

The probability of selecting exactly 2 red and 3 green balls in 5 draws is as follows ($n = 5$, $x = 2$, $n - x = 3$, $p = .4$, $q = .6$):

$$y = \frac{n!}{x!\,(n-x)!}\, p^x q^{n-x}$$

$$y = \frac{5!}{2!\,3!}\,(.4)^2\,(.6)^3 = \frac{(5)(4)(3)(2)(1)}{(2)(1)(3)(2)(1)}\,(.16)\,(.216) = .34560$$

Table 6.3 illustrates the computation of probabilities for all possible numbers of red and green balls which may be drawn from the 10 balls in the bowl. Each ball is replaced in the bowl as soon as it is selected and identified as to color. The sum of the probabilities equals 1.00000.

TABLE 6.3 BINOMIAL PROBABILITY DISTRIBUTION: EXPERIMENT INDICATING COMPUTATION OF PROBABILITIES IN DRAWING SPECIFIED NUMBERS OF RED AND GREEN BALLS FROM A BOWL CONTAINING 4 RED AND 6 GREEN BALLS
(Each ball is replaced in the bowl as soon as it is selected and identified as to color.)

Number of balls to be drawn			
Red	Green	Computation of probability	Probability
5	0	$y = \frac{5!}{5!\,0!}\,(.4)^5\,(.6)^0 =$	.01024
4	1	$y = \frac{5!}{4!\,1!}\,(.4)^4\,(.6)^1 =$	.07680
3	2	$y = \frac{5!}{3!\,2!}\,(.4)^3\,(.6)^2 =$	.23040
2	3	$y = \frac{5!}{2!\,3!}\,(.4)^2\,(.6)^3 =$	.34560
1	4	$y = \frac{5!}{1!\,4!}\,(.4)^1\,(.6)^4 =$	.25920
0	5	$y = \frac{5!}{0!\,5!}\,(.4)^0\,(.6)^5 =$	.07776
		Sum of the probabilities =	1.00000

The probability for each combination of red and green balls can be checked by expansion of the binomial $(p + q)^n$ as follows (Table 6.4):

$$y = (p + q)^5 = p^5 + 5p^4q + 10p^3q^2 + 10p^2q^3 + 5pq^4 + q^5$$

TABLE 6.4 BINOMIAL PROBABILITY DISTRIBUTION: A CHECK ON COMPUTATION OF PROBABILITIES DERIVED IN TABLE 6.3

Number of balls to be drawn			
Red	Green	Computation of probability	Probability
5	0	$y = p^5 = (.4)^5$	= .01024
4	1	$y = 5p^4q = 5(.4)^4(.6) = 5(.0256)(.6)$	= .07680
3	2	$y = 10p^3q^2 = 10(.4)^3(.6)^2 = 10(.064)(.36)$	= .23040
2	3	$y = 10p^2q^3 = 10(.4)^2(.6)^3 = 10(.16)(.216)$	= .34560
1	4	$y = 5pq^4 = 5(.4)(.6)^4 = 5(.4)(.1296)$	= .25920
0	5	$y = q^5 = (.6)^5$	= .07776
		Sum of the probabilities	= 1.00000

6.9 CHARACTERISTICS OF THE BINOMIAL DISTRIBUTION

1. It is a discrete distribution, one with a discrete variable.

DEFINITION: A *discrete variable* is one which can assume a finite number of values in a given interval. They are usually counts and are whole numbers, or integers, such as: persons, family size, houses, ways of obtaining required points in dice tossing.

A discrete variable may be contrasted with a continuous variable.

A *continuous variable* is one which can assume an infinite number of values in a given interval. They are usually measurements which are generally fractional values occurring at minute intervals, such as: heights, weights, and temperatures.

Because the data of Table 6.2 are discrete, Figure 6.1 is drawn as a set of vertical lines, the height of each line representing the probability that the variable x, which is the point number total on two dice, will occur. The vertical lines on the graph merely serve to guide the eye. The actual graph consists of just the eleven points indicated by the dots at the tops of the lines.

2. Its form depends on the values of p and n.
3. The distribution will be symmetrical if $p = q$, and asymmetrical if

$p \neq q$. (The symbol $\neq$ means "does not equal.") If $p \neq q$, then, as n, or the number of trials, increases, the degree of skewness decreases sharply.

4. The mean of a binomial distribution equals np, and the standard deviation equals $\sqrt{npq}$.

Although the problems used to introduce the binomial distribution are based upon games of chance, this distribution is very useful for solving a variety of practical problems, such as: opinion polling and statistical quality control problems.

6.10 THE NORMAL DISTRIBUTION

The Normal Distribution Is the Limit of the Binomial

As the value of n approaches infinity in the expansion of the binomial $(p + q)^n$, a smoother and more symmetrical curve is developed. Thus, as the number of trials increases, the binomial distribution becomes more and more like the normal distribution in appearance. The normal distribution, therefore, may be thought of as the limit of the binomial distribution.

The Normal Curve

A normal distribution, illustrated in Figure 6.3, is called the *normal curve of error* because it is attributed to the laws of chance. The symbol μ is the small Greek letter mu which is used to designate the arithmetic mean in a universe or population. The symbol σ is the small Greek letter sigma, which is used to designate the standard deviation of a universe or population.

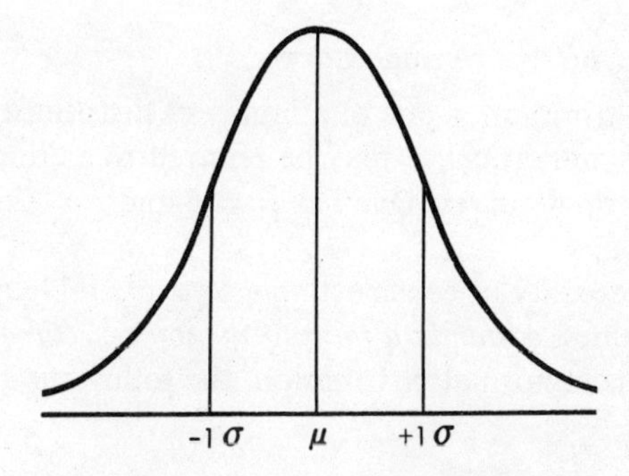

Fig. 6.3 The Normal Curve

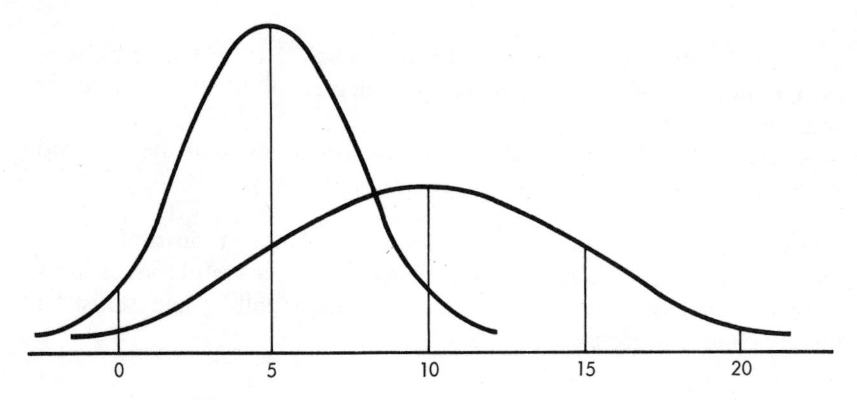

Fig. 6.4 Two Examples of the Normal Curve

The normal distribution is a continuous curve whose mathematical characteristics were defined in the eighteenth century by Abraham De Moivre (1667–1745). Subsequently, it was independently discovered early in the nineteenth century by Pierre S. Laplace (1749–1827) and Carl F. Gauss (1777–1855). The normal distribution is also known, therefore, as the *Gaussian* or *Laplacian* distribution.

The normal curve may be completely defined if its arithmetic mean and standard deviation are known. If these two values are available, the mathematical equation for the normal curve permits calculation of the *height* of the curve for each point as well as the *area* under the curve between any two points on the X axis. Actually, a table of ordinates (heights) and a table of areas have been prepared and are available for use.

Of all the statistical tables included at the back of this book, the one with the most relevance to the subject of the normal curve is the table of areas (Appendix Table III). This table provides the probabilities of particular values falling into specified intervals.

Standard Form of the Normal Curve

The normal distribution is one of a family of distributions. If a curve is bell-shaped and symmetrical, it may be referred to as normal. Figure 6.4 illustrates two normal curves. One has $\mu = 5$ and $\sigma = 2.5$. The other has $\mu = 10$ and $\sigma = 5$.

To avoid the necessity of constructing a separate table of areas for each set of μ and σ values, a *standard form of the normal curve* was developed. For the standardized normal distribution, the following applies:

1. $\mu = 0$ and $\sigma = 1$
2. The area under the curve $= 1$

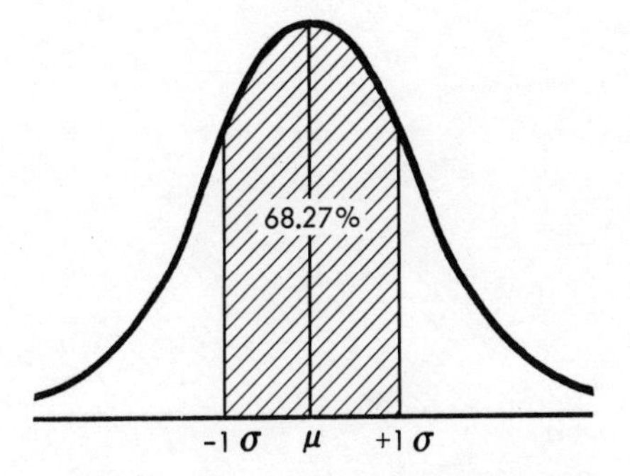

Fig. 6.5 Area of the Standard Form of the Normal Curve between $+1\sigma$ and -1σ

The area under the normal distribution between $\mu \pm 1\sigma$ is 0.6827 out of a total area of 1.

Relationship Between Arithmetic Mean and Standard Deviation in a Population

In the standard form of the normal curve, the following relationships apply:

$\mu \pm 1\sigma$ contains 68.27% of the area under the curve.
$\mu \pm 2\sigma$ contains 95.45% of the area under the curve.
$\mu \pm 3\sigma$ contains 99.73% of the area under the curve.

Figures 6.5–6.7 illustrate these relationships.

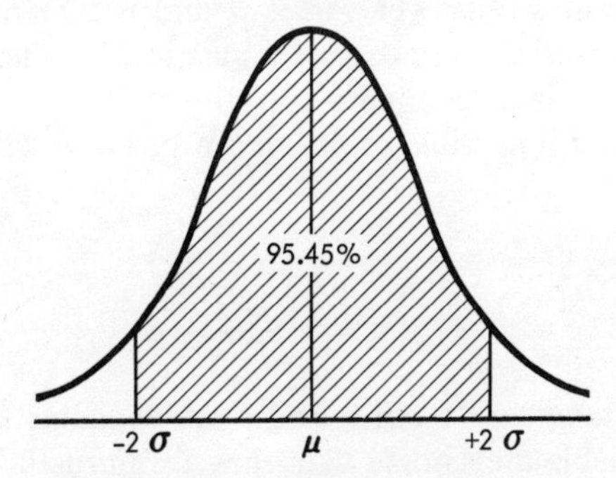

Fig. 6.6 Area of the Standard Form of the Normal Curve between $+2\sigma$ and -2σ

The area under the normal distribution between $\mu \pm 2\sigma$ is 0.9545 out of a total area of 1.

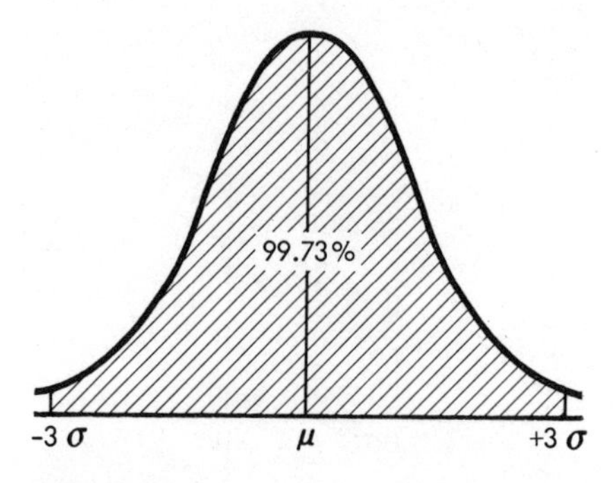

Fig. 6.7 Area of the Standard Form of the Normal Curve between $+3\sigma$ and -3σ

The area under the normal distribution between $\mu \pm 3\sigma$ is 0.9973 out of a total area of 1.

Figure 6.7 clearly indicates that the amount of area under the curve less than -3σ or more than $+3\sigma$ is negligible, .0027 of a total area of 1. Both tails actually extend from $-\infty$ (infinity) to $+\infty$.

6.11 THE *z* TRANSFORMATION

Any normal curve can be transformed into the standard normal curve by changing its scale. Whatever values μ and σ take on the X scale, μ is set to equal 0, and σ is set to equal 1 on a z scale of the standard normal curve. This is illustrated by Figure 6.8.

A formula makes it possible to change from the x scale to the z scale, and vice versa.

FORMULA:

$$z = \frac{X - \mu}{\sigma}$$

The above formula is used when μ and σ are available for a population. If all that is available are sample data that are normally distributed, the following formula is used:

$$z = \frac{X - \bar{x}}{s}$$

Both formulas serve in precisely the same way. Once the z value is obtained, areas under the normal curve are determined from statistical Appendix

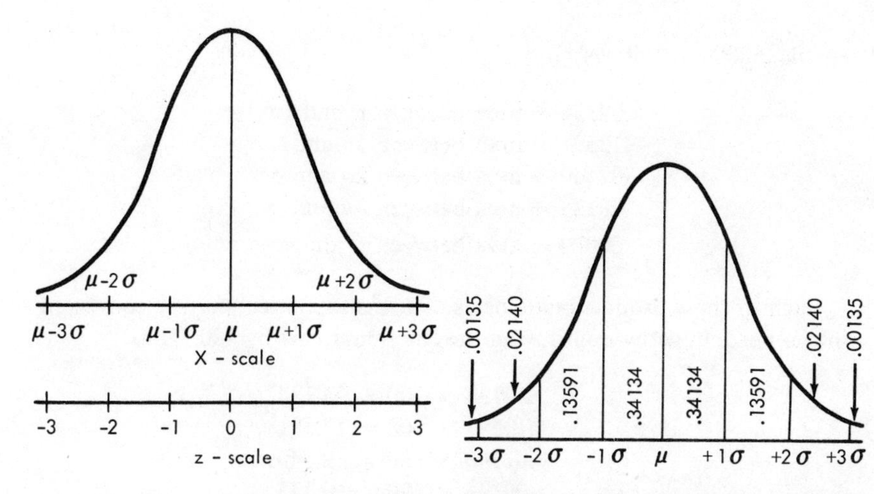

Fig. 6.8 The z and X Scales of the Standard Normal Curve

Fig. 6.9 Areas under the Normal Curve between Perpendiculars Erected at Each σ Distance from μ

Table III. This table provides probabilities of values falling into given areas. These probabilities are equivalent to the areas under the curve.

EXAMPLE 1: If computed z values were precisely equal to $\pm 1\sigma$, $\pm 2\sigma$, and $\pm 3\sigma$ from μ, what is the area of the normal curve included between each perpendicular erected at each z or σ distance from μ?

SOLUTION: From Appendix Table III the area under the curve is ascertained at 1σ, 2σ, and 3σ from μ.

$$\text{The area between } \mu \text{ and } 1\sigma = .34134$$
$$\text{The area between } \mu \text{ and } 2\sigma = .47725$$
$$\text{The area between } \mu \text{ and } 3\sigma = .49865$$

The area between ordinates is derived by subtraction.

$$\text{The area between } +1\sigma \text{ and } +2\sigma = .47725 - .34134 = .13591$$
$$\text{The area between } +2\sigma \text{ and } +3\sigma = .49865 - .47725 = .02140$$
$$\text{The area between } +3\sigma \text{ and } \infty = .50000 - .49865 = .00135$$

Since the normal curve is symmetrical around the mean, equal distances between perpendiculars erected at each σ distance from μ are the same for each half of the normal curve. This is illustrated in Figure 6.9.

The sum of the individual portions for each half of the normal curve equals .50000 as follows:

$$\begin{aligned}
.34134 &= \text{area between } \mu \text{ and } 1\sigma \\
.13591 &= \text{area between } 1\sigma \text{ and } 2\sigma \\
.02140 &= \text{area between } 2\sigma \text{ and } 3\sigma \\
.00135 &= \text{area between } 3\sigma \text{ and } \infty \\
\hline
.50000 &= \text{area between } \mu \text{ and } \infty
\end{aligned}$$

Each of these proportionate parts of 1, the total area, may be converted into a percentage by multiplying the fractional part by 100. Thus,

$$\begin{aligned}
.34134 \times 100 &= 34.134\% \\
.13591 \times 100 &= 13.591 \\
.02140 \times 100 &= 2.140 \\
.00135 \times 100 &= 0.135 \\
\text{Total} = .50000 \times 100 &= 50.000\%
\end{aligned}$$

EXAMPLE 2: The frequency distribution in Table 6.5 represents a sample of 120 dry cell batteries distributed according to length of life. If $\bar{x} = 11.35$ hours and $s = 3.03$ hours, and assuming that the data are normally distributed, what is the probability that a battery selected at random will have a life of: (a) less than 8 hours? (b) more than 16 hours? (c) between 9 and 17 hours?

TABLE 6.5 LIFE SPAN OF 120 DRY CELL BATTERIES

Life in hours (l_1–l_2)	X	Number of batteries (f)
2– 3.99		1
4– 5.99		3
6– 7.99		10
8– 9.99		21
10–11.99		42
12–13.99		24
14–15.99		10
16–17.99		6
18–19.99		2
20–21.99		1
		$n = 120$

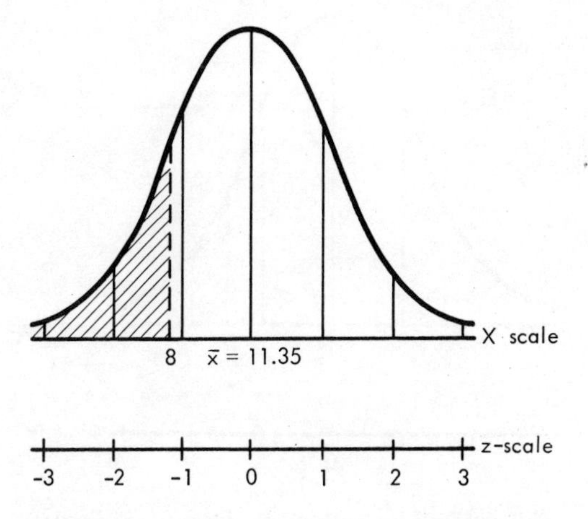

Fig. 6.10 Illustration of Use of z Transformation

SOLUTION: a. Less than 8 hours

$$z = \frac{X - \bar{x}}{s}$$

$$= \frac{8 - 11.35}{3.03}$$

$$= \frac{-3.35}{3.03}$$

$$= -1.1056 \quad \text{or} \quad -1.11$$

According to Appendix Table III, Areas Under the Normal Curve, a z value of 1.11 indicates an area of .36650 between $\bar{x}$, or μ, and $-1.11z$. This is illustrated in Figure 6.10.

The probability that a battery will burn less than 8 hours is indicated by the shaded area of the normal curve.

The total area of the left half of the curve	= .50000
The area of the curve from $\bar{x}$ to $-1.11z$	= .36650
Therefore, by subtraction, the area from $-1.11z$ to $-\infty$	= .13350 × 100
	= 13.4%

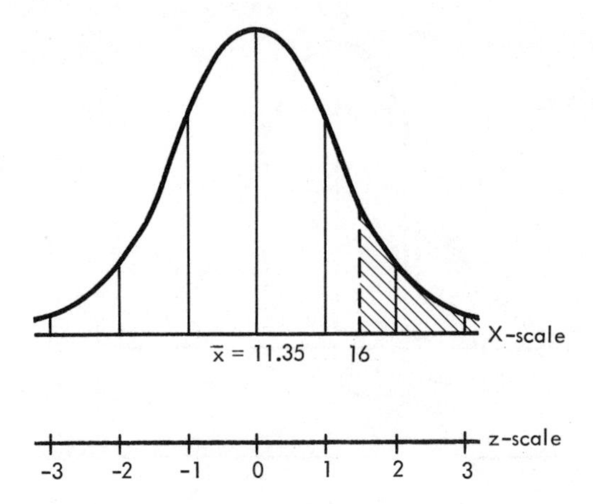

Fig. 6.11 Illustration of Use of z Transformation

Thus, the probability that a dry cell battery selected at random will have a life of less than 8 hours is 13.4 per cent. Or, the probability is that about 13 batteries out of 100 will have a life of less than 8 hours.

b. More than 16 hours

$$z = \frac{X - \bar{x}}{s}$$

$$= \frac{16 - 11.35}{3.03}$$

$$= \frac{4.65}{3.03}$$

$$= 1.5347, \quad \text{or} \quad 1.53$$

According to Table III, a z value of 1.53 indicates an area of .43699 between $\bar{x}$, or μ, and $1.53z$. This is illustrated in Figure 6.11. The probability that a battery will burn more than 16 hours is indicated by the shaded area of the normal curve.

The total area of the right half of the curve = .50000
The area of the curve from $\bar{x}$ to $1.53z$ = .43699
Therefore, by subtraction, the area from $1.53z$ to $+\infty$ = .06301 × 100
= 6.3%

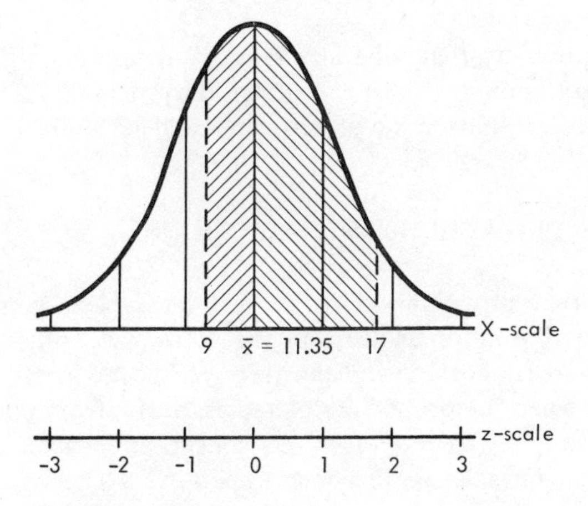

Fig. 6.12 Illustration of Use of z Transformation

Thus, the probability that a battery selected at random will have a life of more than 16 hours is 6.3 per cent. Or, the probability is that about 6 bulbs out of 100 will have a life of more than 16 hours.

c. Between 9 and 17 hours

$$z = \frac{X - \bar{x}}{s} = \frac{9 - 11.35}{3.03} = \frac{-2.35}{3.03} = -.7756 \quad \text{or} \quad -.78$$

$$z = \frac{X - \bar{x}}{s} = \frac{17 - 11.35}{3.03} = \frac{5.65}{3.03} = 1.8647 \quad \text{or} \quad 1.86$$

Appendix Table III, indicates an area of .28230 for $0.78z$, and an area of .46856 for $1.86z$. This is illustrated in Figure 6.12. The probability that a battery will burn between 9 and 17 hours is indicated by the shaded area of the normal curve.

The area of the curve from $\bar{x}$ to $-0.78z$ = .28230
The area of the curve from $\bar{x}$ to $+1.86z$ = +.46856
By addition, the area from $-0.78z$ to $+1.86z$ = .75086 × 100
= 75.1%

Thus, the probability that a battery selected at random will have a life between 9 and 17 hours is 75.1 per cent. Or, the probability is that about 75 batteries out of 100 have a life of between 9 and 17 hours.

6.12 OTHER PROBABILITY DISTRIBUTIONS

Thus far, two probability distributions have been considered—the binomial and the normal distributions. However, other probability distributions are important in statistics. When the trials are not independent or when the probability of success varies from one trial to the next, the binomial distribution is no longer applicable. Instead, the *hypergeometric distribution* becomes operative. The hypergeometric distribution is used when the probability of success changes from trial to trial. For example, one of the problems used to illustrate the binomial distribution on page 95 assumes that each ball is *replaced* in the bowl as soon as it is selected and identified as to color. However, if after each draw, the ball is *not replaced*, the probability of selecting a particular colored ball changes. If the experiment starts with 4 red and 6 green balls, after the first ball is drawn, only 9 balls remain, 3 red and 6 green, or 4 red and 5 green, and so forth.

The Poisson probability distribution was developed by Simeon D. Poisson (1781–1840), in 1837. It is useful in the solution of many sampling problems, particularly when the number of independent events is large, and when for each independent event the probability that a particular outcome will result is small. Thus, a Poisson distribution may be expected in problems where the chance of any individual event being a success is small.

As an example, the *Poisson distribution* is widely applied in statistical quality control work. If an item is being mass-produced, that is, the number of items (n) is large, and if the proportion defective in any lot is expected to be very small, that is, the probability of defective items (p) is small, then the *Poisson probability distribution* may be used.

In other statistical problems, a number of continuous probability distributions are useful, in addition to the *normal curve*. These include the *t distribution*, the *chi-square* (χ^2) *distribution*, and the F distribution. The t, χ^2, and F probability distributions assume that the population from which a sample is drawn is normally distributed. For this reason, these distributions are called *derived* distributions. The t and χ^2 distributions will be considered in the following chapter.

SUMMARY: Common statistical probability distributions include the following:

Discrete	*Continuous*
1. binomial	1. normal
2. hypergeometric	2. t
3. Poisson	3. χ^2
	4. F

EXERCISES

1. Define or explain the following:

A priori probability
Addition rule of probability
Area under the normal curve
Bell-shaped curve
Binomial
Binomial coefficient
Binomial expansion
Binomial probability expansion
Binomial theorem
Combination
Continuous probability distribution
Continuous variable
Discrete probability distribution
Discrete variable
Empirical probability
Factorial
Gaussian distribution
Hypergeometric distribution
Independent events
Joint occurrence
Laplacian distribution
Laws of chance
Limit of the binomial
Multiplication rule of independent events
Mutually exclusive events
Pascal's triangle
Permutation
Poisson probability distribution
Probability
Probability distribution
Standard form of the normal curve
Table of areas
Table of ordinates
z scale
z transformation

2. In tossing two dice, what is the probability of

a. A sum of 8?
b. A sum of 11?
c. A sum of 6 or less?
d. A sum of 9 or more?

3. Consider the possible outcomes of a single toss of four coins.

   a. Complete Table 6.6. Label each possible outcome, as the case may be, under the appropriate column. For example, if $T =$ tail and $H =$ head, then $THTH$ is one possible outcome.

TABLE 6.6 POSSIBLE OUTCOMES OF A SINGLE TOSS OF FOUR COINS

0 Heads	1 Head	2 Heads	3 Heads	4 Heads
TTTT	TTTH	TTHH	THHH	HHHH
	TTHT	THHT	HHHT	
	THTT	HHTT	HHTH	
	HTTT	HTTH	HTHH	
		HTHT		
		THTH		

   b. Using the binomial expansion, compute the following probabilities.
      (1) P (0 heads)
      (2) P (1 head)
      (3) P (2 heads)
      (4) P (3 heads)
      (5) P (4 heads)

   c. Draw the distribution of probabilities derived in b above, using Figure 6.1, page 89, as the model.

4. The output of a manufacturing firm producing metal stampings is 60 per cent top grade and 40 per cent average grade.

   a. In drawing a sample of two metal stampings, what is the probability of drawing one top grade and one average grade?

   b. In drawing a sample of three metal stampings, what is the probability of drawing two top grade and one average grade?

5. Given a population which is 70 per cent top grade and 30 per cent average grade;

   a. What is the probability of drawing 2 average-grade items in a sample of 3?

   b. What is the probability of drawing 3 average-grade items in a sample of 4?

6. The frequency distribution in Table 6.7 represents units produced by a sample of 50 workers.

TABLE 6.7 DAILY OUTPUT OF A PRODUCT

Units produced	Number of workers
75–124	6
125–174	8
175–224	14
225–274	11
275–324	7
325–374	4
Total	50

Assuming that the data are normally distributed, if $\bar{x} = 217$ and $s = 71$, what is the probability that a worker selected at random will produce

a. Less than 155 units?
b. More than 305 units?
c. Between 165 and 298 units?

7

Statistical Inference and Hypothesis Testing

This chapter on statistical inference and hypothesis testing considers first the sampling process, the types of survey coverage, and the major types of samples. It then analyzes the use of a table of random numbers to insure the selection of a probability sample.

It is then shown why the sampling distribution of the arithmetic mean makes it possible to generalize about a population on the basis of data derived from a single random sample. The central limit theorem, which follows, explains why statistics developed from a random sample approach the normal probability distribution as the sample size is increased.

The standard error of the mean becomes the bridge between a sample and population by providing the probability limits within which the true mean of the population is located, and which are based upon measures derived from a single random sample. The concept of degrees of freedom is then considered to explain particularly the problems of inference for small samples.

Statistical inference is the process of reaching a conclusion about a population by statistical measurements of a randomly selected sample. A discussion of the types of inference problems follows in this chapter. These are illustrated by examples in interval estimation for a large random sample as well as a small random sample which utilizes the t *distribution.*

The chapter then analyzes hypothesis testing, the two types of errors which may be made, the general types of problems encountered, and detailed solutions to illustrative cases.

A discussion of the factors determining sample size follows, and the chapter concludes with the chi-square probability distribution, which is illustrated by problems testing the goodness of fit.

7.1 THE SAMPLING PROCESS

Sampling becomes necessary when it is too costly or too time-consuming to examine the entire population. In such a situation, a part of the totality is selected and studied intensively, and then conclusions are made about the totality from which the sample was drawn. Various types of coverage of the entire population are available to the statistician.

7.2 TYPES OF SURVEY COVERAGE

Three types of survey coverage are possible.

Universal coverage. In universal coverage every element of the population is included in the survey. For example, the decennial census of the population of the United States taken by the U.S. Department of Commerce, Bureau of the Census, is a complete enumeration of each person in the population.

Sample coverage. Because of the time and expense involved in a census-type approach to collecting information, the statistician resorts to a sampling process. He collects data for only a part of the population. After a study of the sample, the statistician draws conclusions about the population.

Combination of universal and sample coverage. The decennial census of housing taken by the U.S. Department of Commerce, Bureau of the Census, is a combination of universal, or total, coverage and sample coverage. Every household is enumerated for certain basic housing information, and in addition every fourth household, a 25 per cent sample, is asked to answer specified supplementary questions.

7.3 THE MAJOR TYPES OF SAMPLES

Random Sample

In a random sample each item in the population has an equal chance of being selected. The laws of chance determine which item is included in a sample selected at random, thus making it possible to calculate probabilities.

FORMS OF THE RANDOM SAMPLE:

A systematic sample is one in which, for example, every tenth item is selected after a population is arrayed in some form, and after a random starting point is set.

Double sampling. If after the first sample of n items is drawn, analysis of the data is inconclusive, another sample must be selected to be added to the first sample, thus strengthening it.

Sequential sampling. This is a variation of double sampling in that the additional sample need not be the same size as the original sample.

Area sampling. A geographical region is subdivided into subareas. A random sample is then taken of the subareas or a portion of them. For example, sampling of families in a city may be achieved by using city blocks as the subareas.

Cluster sampling. A cluster is a grouping of individuals, families, or houses, which may be randomly selected within a city block. Cluster sampling, used primarily in conjunction with area sampling, saves travel and interviewing time, and results in cost savings.

Purposive Sample

The purposive sample contrasts directly with the probability or random sample. The purposive sample is selected by design, or to meet the needs of a particular objective. It is based upon the statistician's prior knowledge of the characteristics of the population from which the sample is selected.

FORMS OF THE PURPOSIVE SAMPLE:

Judgment sample. In this type of sample, judgment plays a substantial role either in selecting the sample items or in making decisions about large parts of the population for which the sample provides no information.

The quota sample is another form of the judgment sample. In a quota sample, the interviewer is required to question a specified number of persons with given characteristics. The quotas are generally set up according to certain known characteristics of the population, such as age, sex, and income. The interviewer may select the persons he will interview as long as they meet the specifications.

LIMITATION OF THE PURPOSIVE SAMPLE: Because it is difficult to obtain agreement on the representativeness of the purposive sample, the results of a study based upon this type of sample do not lend themselves to probability analysis.

7.4 ADDITIONAL CONSIDERATIONS IN SAMPLE DESIGN

Homogeneous Data vs. Heterogeneous Data

Knowledge of the population or universe is important in designing a sample. Inferences about a totality from which a sample is drawn can be valid only if the sample is representative of that totality. For example, if an investigation is concerned about the attitude of college students in general toward a particular problem, then all college students taken together may be considered a *homogeneous* population. The attitude to be studied is treated as a uniform characteristic in the totality. On the other hand, if the attitude to be investigated is toward a particular draft law as it relates to college student service, then the population must be treated as *heterogeneous*, because it may then be divided into a number of subgroups. The attitude will differ by age, by sex, and by year in college, and perhaps by draft status.

Stratified Sampling

Heterogeneous data may be subdivided into homogeneous data, *strata*, or classes. Each class or *stratum* will represent a particular element in the population. Each stratum is basically a separate population, and is sampled independently. The relative importance of each stratum determines its weight in the total sample, thus permitting aggregation of the parts to yield the total. Therefore, a *stratified sample* is a combination of independent samples selected in carefully computed proportions from homogeneous strata within a heterogeneous population.

7.5 USING A TABLE OF RANDOM NUMBERS

A statistically random selection of items for a sample will yield a *probability sample*, i.e., all items in the population have preassigned and known probabilities of being chosen. Such a random selection of items may be readily accomplished with the aid of a *table of random digits*. Appendix Table IV is part of one such table. This table may be considered as having been constructed by some random process, such as throwing a pair of dice over and over or drawing colored balls out of a bowl. In practice, a much more complicated process is used, involving complex mathematical and technical details.

The table of random digits represents a thorough scrambling of the digits 0, 1, 2, 3, 4, 5, 6, 7, 8, and 9. No matter how these digits are read on such a table, down, up, or sideways, they maintain their characteristic of being random. Every digit is independent of the preceding one or of the succeeding one.

A number of tables of random numbers are available. Two of the leading tables are the Interstate Commerce Commission's *Table of 105,000 Random Decimal Digits*, published in 1949, and *A Million Random Digits with 100,000 Normal Deviates*, prepared by the Rand Corporation and published in 1956.

EXAMPLE: Use the table of random digits to select 250 clerks from a numbered list of 8,200 clerks.

SOLUTION: The I.C.C. table of random digits contains thirty pages. Any page may be selected, and it is permissible to begin at any point on that page. Four-digit numbers are required, because the problem specifies the population size as 8,200. This number consists of four digits. If three-digit numbers were used, clerks on the numbered list larger than 100 would be excluded from the sample. If five-digit numbers were used, too many numbers would be drawn for which no equivalent clerk would be listed in the population.

Using Appendix Table IV assume that column 1, row 11, is selected as the starting point. Using four digits, the first item in the sample is 2891. The second item is 6355, third 942, fourth 1036, fifth 711, sixth 5108, seventh 236, and so forth. The twentieth item, 9192, would be excluded, because no such number appears in the population. This number therefore would be passed over, and the next one would be selected. If the same number appears a second or a third time, pass over it, and move on to the next.

After column 1 is used up, column 2 is started at the top, using either the first four digits as in column 1, or combining the fifth digit of column 1 with the first 3 digits of column 2 to make the required four-digit numbers.

In this manner, 250 numbers are selected, and the equivalent-numbered clerks listed in the population would make up the 250-item sample.

7.6 SAMPLING DISTRIBUTION OF THE ARITHMETIC MEAN

THE CONCEPT: The *sampling distribution of the arithmetic mean* is one of the basic concepts of statistical inference. It provides an explanation of the reason why it is possible to generalize about a population by using information derived from a single random sample.

The concept of the sampling distribution of the arithmetic mean is based upon a single population and a great many random samples drawn from this population. The arithmetic mean is computed for each random sample drawn from this population. A frequency distribution is prepared of the sample arithmetic means. Then, a mean of sample means and standard deviation of sample means are calculated. The relationship among the computed values is interesting.

1. The arithmetic mean of the sample means is almost identical with the arithmetic mean of the population.

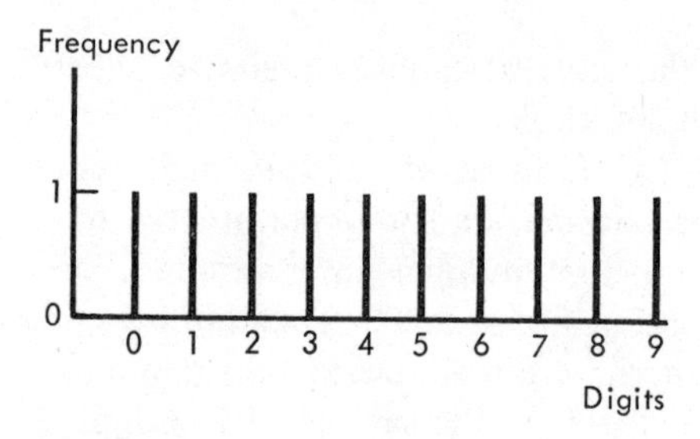

Fig. 7.1 Distribution of a Population of 10 Digits

2. The standard deviation of the sampling distribution of sampling means is smaller than the standard deviation of the population.

DEFINITION: The *sampling distribution of the arithmetic mean* is a frequency distribution of the arithmetic means of an infinite number of large random samples drawn from a population. The assumptions made in this definition are:

1. The samples drawn are random, and are therefore subject to the laws of probability.
2. The sample size n is large. A large sample should consist of 30 or more items.
3. The number of random samples drawn is infinite. This is a theoretical requirement.

Actually, sampling distributions are abstractions, but their characteristics are known.

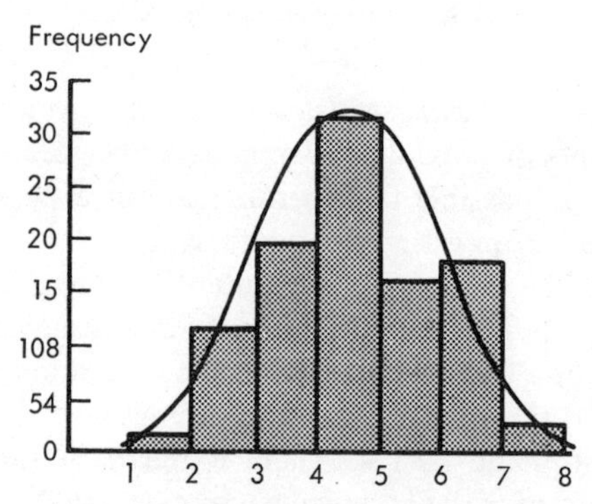

Fig. 7.2 Sampling Distribution of Arithmetic Means of 100 Random Samples of 5 Digits Each
SOURCE: Table 7.2.

EXAMPLE: Because an understanding of the relationship between the mean of the population and the sampling distribution and between the standard deviation of the population and the sampling distribution are essential to an understanding of statistical inference, a simple experiment may be conducted to illustrate these results.

1. Compute the arithmetic mean and standard deviation for the population of ten digits: 0, 1, 2, 3, 4, 5, 6, 7, 8, and 9.
2. Draw 100 random samples of 5 digits each, using a table of random numbers.
3. Compute the arithmetic mean of each sample.
4. Prepare a frequency distribution for these 100 sample means.
5. Compute the arithmetic mean for the 100 sample means.
6. Compute the standard deviation for the 100 sample means.
7. Compare the arithmetic mean and standard deviation of the population with the arithmetic mean and standard deviation of the *sampling distribution of the arithmetic mean.*

This example provides merely an empirical approximation of the theoretical sampling distribution because, first, the size of each random sample is 5, and not 30 or more; and, second, 100 random samples are drawn, and not an infinite number. Nevertheless, it illustrates the principle.

SOLUTION:

1. The arithmetic mean of the population is 4.5, and the standard deviation is 2.8723. These were computed as follows:

$$\mu = \frac{\Sigma X}{N}$$

$$\mu = \frac{45}{10}$$

$$\mu = 4.5$$

and

$$\sigma = \sqrt{\frac{\Sigma (X - \mu)^2}{N}}$$

$$\sigma = \sqrt{\frac{82.50}{10}}$$

$$\sigma = \sqrt{8.250}$$

$$\sigma = 2.8723$$

2 and 3. In Table 7.1 are 100 random samples of 5 digits each selected from this population. A table of random digits represents the population.

TABLE 7.1 100 RANDOM SAMPLES OF 5 DIGITS EACH AND THE ARITHMETIC MEAN OF EACH SAMPLE

Sample number	Random sample					Sum of the items	Arithmetic mean
1	1	0	0	9	7	17	3.4
2	3	7	5	4	2	21	4.2
3	0	8	4	2	2	16	3.2
4	9	9	0	1	9	28	5.6
5	1	2	8	0	7	18	3.6
6	6	6	0	6	5	23	4.6
7	3	1	0	6	0	10	2.0
8	8	5	2	6	9	30	6.0
9	6	3	5	7	3	24	4.8
10	7	3	7	9	6	32	6.4
11	9	8	5	2	0	24	4.8
12	1	1	8	0	5	15	3.0
13	8	3	4	5	2	22	4.4
14	8	8	6	8	5	35	7.0
15	9	9	5	9	4	36	7.2
16	6	5	4	8	1	24	4.8
17	8	0	1	2	4	15	3.0
18	7	4	3	5	0	19	3.8
19	6	9	9	1	6	31	6.2
20	0	9	8	9	3	29	5.8
21	9	1	4	9	9	32	6.4
22	8	0	3	3	6	20	4.0
23	4	4	1	0	4	13	2.6
24	1	2	5	5	0	13	2.6
25	6	3	6	0	6	21	4.2
26	6	1	1	9	6	23	4.6
27	1	5	4	7	4	21	4.2
28	9	4	5	5	7	30	6.0
29	4	2	4	8	1	19	3.8
30	2	3	5	2	3	15	3.0
31	0	4	4	9	3	20	4.0
32	0	0	5	4	9	18	3.6
33	3	5	9	6	3	26	5.2
34	5	9	8	0	8	30	6.0
35	4	6	0	5	8	23	4.6
36	3	2	1	7	9	22	4.4
37	6	9	2	3	4	24	4.8
38	1	9	5	6	5	26	5.2
39	4	5	1	5	5	20	4.0
40	9	4	8	6	4	31	6.2
41	9	8	0	8	6	31	6.2
42	3	3	1	8	5	20	4.0
43	8	0	9	5	1	23	4.6
44	7	9	7	5	2	30	6.0
45	1	8	6	3	3	21	4.2
46	7	4	0	2	9	22	4.4
47	5	4	1	7	8	25	5.0
48	1	1	6	6	4	18	3.6
49	4	8	3	2	4	21	4.2
50	6	9	0	7	4	26	5.2

(Table 7.1 continued)

TABLE 7.1 (CONTINUED)

Sample number	Random sample					Sum of the items	Arithmetic mean
51	3	2	5	3	3	16	3.2
52	0	4	8	0	5	17	3.4
53	6	8	9	5	3	31	6.2
54	0	2	5	2	9	18	3.6
55	9	9	9	7	0	34	6.8
56	7	4	7	1	7	26	5.2
57	1	0	8	0	5	14	2.8
58	7	7	6	0	2	22	4.4
59	3	2	1	3	5	14	2.8
60	4	5	7	5	3	24	4.8
61	1	7	7	6	7	28	5.6
62	0	5	4	3	1	13	2.6
63	9	9	6	3	4	31	6.2
64	4	0	2	0	0	6	1.2
65	6	7	3	4	8	28	5.6
66	1	7	6	7	4	25	5.0
67	3	5	6	3	5	22	4.4
68	9	9	8	1	7	34	6.8
69	2	6	8	0	3	19	3.8
70	2	0	5	0	5	12	2.4
71	1	4	5	2	3	15	3.0
72	9	4	5	9	8	35	7.0
73	8	1	9	4	9	31	6.2
74	7	3	7	4	2	23	4.6
75	4	9	3	2	9	27	5.4
76	9	0	4	4	6	23	4.6
77	4	5	2	6	6	23	4.6
78	2	8	5	7	3	25	5.0
79	1	6	2	1	3	13	2.6
80	7	8	3	1	7	26	5.2
81	5	2	4	9	4	24	4.8
82	9	7	6	5	4	31	6.2
83	1	5	3	0	7	16	3.2
84	0	8	3	9	1	21	4.2
85	8	5	2	3	6	24	4.8
86	0	0	5	9	7	21	4.2
87	6	1	4	0	6	17	3.4
88	4	1	4	3	0	12	2.4
89	1	4	9	3	8	25	5.0
90	3	1	9	9	4	26	5.2
91	2	4	8	2	6	22	4.4
92	1	6	2	3	2	14	2.8
93	0	0	4	0	6	10	2.0
94	4	9	1	4	0	18	3.6
95	3	2	5	3	7	20	4.0
96	4	3	9	0	2	18	3.6
97	4	5	6	1	1	17	3.4
98	4	9	8	8	3	32	6.4
99	7	7	9	2	8	33	6.6
100	9	4	1	3	8	25	5.0

4. The frequency distribution of the 100 sample arithmetic means is shown in Table 7.2.

TABLE 7.2 FREQUENCY DISTRIBUTION OF ARITHMETIC MEANS OF 100 RANDOM SAMPLES OF 5 DIGITS EACH

l_1–l_2	Tally	f
1.0–1.9	/	1
2.0–2.9	~~////~~ ~~////~~ /	11
3.0–3.9	~~////~~ ~~////~~ ~~////~~ ~~////~~	20
4.0–4.9	~~////~~ ~~////~~ ~~////~~ ~~////~~ ~~////~~ ~~////~~ //	32
5.0–5.9	~~////~~ ~~////~~ ~~////~~ /	16
6.0–6.9	~~////~~ ~~////~~ ~~////~~ //	17
7.0–7.9	///	3
		n = 100

SOURCE: Table 7.1.

5 and 6. Computation of the arithmetic mean and the standard deviation of the sampling distribution of the arithmetic mean is indicated in Table 7.3.

TABLE 7.3 COMPUTATION OF ARITHMETIC MEAN AND STANDARD DEVIATION OF THE SAMPLING DISTRIBUTION OF THE ARITHMETIC MEAN

l_1–l_2	X	f	d	fd		fd^2
1.0–1.9	1.5	1	−3	−3		9
2.0–2.9	2.5	11	−2	−22	−45	44
3.0–3.9	3.5	20	−1	−20		20
4.0–4.9	4.5	32	0	0		0
5.0–5.9	5.5	16	+1	16		16
6.0–6.9	6.5	17	+2	34	+59	68
7.0–7.9	7.5	3	+3	9		27
		n = 100		$\Sigma fd = +14$		$\Sigma fd^2 = 184$

SOURCE: Table 7.2.

Computation of the arithmetic mean:

$$\bar{x} = \bar{x}_g + \frac{\Sigma fd}{n} i$$

$$\bar{x} = 4.5 + \frac{14}{100}(1)$$

$$\bar{x} = 4.5 + .14$$

$$\bar{x} = 4.64$$

Computation of the standard deviation:

$$s = i\sqrt{\frac{\Sigma fd^2}{n} - \left(\frac{\Sigma fd}{n}\right)^2}$$

$$s = 1\sqrt{\frac{184}{100} - \left(\frac{14}{100}\right)^2}$$

$$s = 1\sqrt{1.84 - (.14)^2}$$

$$s = 1\sqrt{1.84 - .0196}$$

$$s = 1\sqrt{1.8204}$$

$$s = 1.3492$$

7. A comparison of the arithmetic mean and the standard deviation of the population with computed measures of the *sampling distribution of the arithmetic mean* is shown in Table 7.4.

TABLE 7.4 COMPARISON OF ARITHMETIC MEAN AND STANDARD DEVIATION OF THE POPULATION AND ARITHMETIC MEAN AND STANDARD DEVIATION OF THE SAMPLING DISTRIBUTION OF 100 RANDOM SAMPLES

Measure	Population	Sampling distribution
Arithmetic mean	4.5	4.64
Standard deviation	2.8723	1.3492

CONCLUSION: The arithmetic means of the two distributions are almost identical. The standard deviation of the sampling distribution of the arithmetic mean is considerably smaller than that of the population.

As illustrated in Figure 7.1, the population of 10 digits has a *rectangular distribution* and yet the sampling distribution of the arithmetic mean with $n = 5$ yields a frequency distribution, shown in Figure 7.2, which resembles the normal distribution. In general, when n is large the sampling distribution of the arithmetic mean will approximate the normal distribution as the number of random samples increases.

7.7 THE CENTRAL LIMIT THEOREM

THE CONCEPT: The *central limit theorem* is often referred to as the most important theorem in statistics. Even for relatively small samples, the tendency of sampling distributions to resemble the normal distribution is a remarkable phenomenon.

The central limit theorem, which has been mathematically proved, generalizes about the impact of the sample size n upon the sampling distribution of the arithmetic mean.

DEFINITION (the *central limit theorem*): As n becomes large, and regardless of the shape of the original population, the sampling distribution of the arithmetic mean will approach the normal curve, assuming the mean and variance exist.

Even though the population of 10 digits is not normally distributed (Figure 7.1), the sampling distribution of the arithmetic mean, computed from samples as small as 5, is a good approximation of the normal curve (Figure 7.2).

APPLICATIONS: The *central limit theorem* permits the statistican to make some important generalizations:

1. A sample mean based upon relatively few observations may be used to approximate the mean of the population from which the sample was drawn.
2. Variations between a sample mean and a population mean may be estimated with a specified degree of certainty.
3. Sample means of repeated random samples of the same size cluster closely around the mean of the population.

7.8 THE STANDARD ERROR OF THE MEAN

THE CONCEPT: The standard deviation of the sampling distribution of the arithmetic mean is called *the standard error of the mean.* It measures the dispersion of sample means around the true population mean. It is also a measure of the probable extent to which a sample mean is likely to vary in future random samples. The value of the standard error of the mean is always smaller than the standard deviation, except when the sample size is 1, in which case the two are equal.

The relationships between the arithmetic mean and standard deviation of the population and the corresponding measures for the sampling distribution have been stated in a theorem which has been proved mathematically.

THEOREM: If random samples of size n are drawn from a population with arithmetic mean μ and standard deviation σ, the sampling distribution of the arithmetic mean $\bar{x}$ has the mean μ and standard deviation $\sigma/\sqrt{n}$.

DEFINITION: The *standard error of the mean* is another name for the standard deviation of the sampling distribution of the arithmetic mean. It provides a probability statement of the limits within which the true mean of the population is located and is based upon information derived from a single sample of n items.

FORMULA:

$$\sigma_{\bar{x}} = \frac{\sigma}{\sqrt{n}}$$

where $\sigma_{\bar{x}}$ = standard error of the mean
σ = standard deviation of the population
n = number of items in the sample

As the sample size, n, increases, variation in a distribution of sample means is reduced. For example, to reduce the variability by 1/2, the sample size must be increased 4 times.

In practical work, the standard deviation, σ, and the arithmetic mean, μ, of the population are unknown. A population is sampled in order to estimate these characteristics of the population. To do this, the *standard error of the mean* is estimated by substituting the standard deviation of a single sample, s, for the standard deviation of the population, σ. The formula thus becomes:

$$\sigma_{\bar{x}} = \frac{s}{\sqrt{n}}$$

EXAMPLE 1: What is the standard error of the mean for the population of digits from 0 to 9, using $\sigma = 2.8723$ and $n = 5$.

Solution:

$$\sigma_{\bar{x}} = \frac{2.8723}{\sqrt{5}}$$

$$\sigma_{\bar{x}} = \frac{2.8723}{2.23}$$

$$\sigma_{\bar{x}} = 1.2880$$

The standard deviation computed from the sampling distribution of the arithmetic mean was 1.3492, shown in Table 7.4, compared with 1.2880, indicated above, a difference of only .0612. This difference may be attributed to sampling error.

It is important to note that the formula for the standard error of the mean is modified for a small sample. This is explained in Section 7.9, Degrees of Freedom, below.

EXAMPLE 2: If the mean and standard deviation of the population are unknown, and only the mean and standard deviation obtained from a single random sample are known, what is the standard error of the mean, given $s = 25$, and the sample size $n = 100$?

Solution:

$$\sigma_{\bar{x}} = \frac{s}{\sqrt{n}}$$

$$\sigma_{\bar{x}} = \frac{25}{\sqrt{100}}$$

$$\sigma_{\bar{x}} = \frac{25}{10}$$

$$\sigma_{\bar{x}} = 2.5$$

7.9 DEGREES OF FREEDOM

THE CONCEPT: The concept of degrees of freedom may be explained as follows: If three numbers, 2, 7, and 9, plus an unknown fourth number, equal 30, then the unknown number must equal 12. One degree of freedom has been lost.

In this illustration, the number of degrees of freedom equals the number of items in the sample, 4, less one degree of freedom which is lost, or 3. That is, $n - 1 = 4 - 1 = 3$.

LARGE VS. SMALL SAMPLES: A sample is large when it contains 30 or more items. A sample is small when it contains less than 30 items. When the sample size, n, is under 30 items, the formula that gives the best estimate of the standard error of the mean is

$$\sigma_{\bar{x}} = \frac{s}{\sqrt{n - 1}}$$

where $\sigma_{\bar{x}}$ = standard error of the mean
s = standard deviation
$n - 1$ = number of items in sample, n, less one

The quantity $n - 1$ is known as the number of *degrees of freedom.*

When sample sizes are large, that is, contain 30 items or more, the formula $\sigma_{\bar{x}} = s/\sqrt{n}$ is used because the impact of $n - 1$ is insignificant.

In computing the *standard deviation* from the sample data (see Chapter 5) the following formulas are used, the choice depending upon the size of the sample:

LARGE SAMPLES	SMALL SAMPLES
A sample is large when it contains 30 or more items.	A sample is small when it contains less than 30 items.

Ungrouped Data

$$s = \sqrt{\frac{\Sigma (x - \bar{x})^2}{n}} \qquad s = \sqrt{\frac{\Sigma (x - \bar{x})^2}{n - 1}}$$

Grouped Data

$$s = i\sqrt{\frac{\Sigma f(d)^2}{n} - \left(\frac{\Sigma fd}{n}\right)^2} \qquad s = i\sqrt{\frac{\Sigma f(d)^2 - \frac{(\Sigma fd)^2}{n}}{n - 1}}$$

In computing the *standard error of the mean*, the following formulas are used, the choice depending upon the size of the sample:

$$\sigma_{\bar{x}} = \frac{s}{\sqrt{n}} \qquad \sigma_{\bar{x}} = \frac{s}{\sqrt{n - 1}}$$

7.10 STATISTICAL INFERENCE

THE CONCEPT: No one can really be completely certain about anything. Yet, each person makes many decisions daily about matters, trivial or important. Consciously or unconsciously probabilities are estimated. For example, an individual starts out in the morning uncertain that he will arrive home safely that evening. He may meet with an accident or even be killed. Nevertheless, the day is planned as though this eventuality is improbable. Actually, probabilities are favorable for a safe return home. Occasionally, however, an event, whose probability is small, may occur, and the individual ends up in a hospital. Because such an event occurs so infrequently, the chances of one more safe return home are virtually certain.

Similarly, a statistical inference may be drawn from a single random sample of size *n*. While we cannot be certain of the variation of a sample mean from the true mean of an unknown population, limits on both sides of the mean can be set for sampling error. While it is not certain that the true mean of the population is included between the two limits, it is certain that the true mean of the population will be included, for example, 90 or 95 per cent of the time.

DEFINITION: *Statistical inference* is the logical process by which, on the basis of a sample, a conclusion is reached about the population from which the sample is drawn.

There are two types of inference problems: *estimation* and *testing hypotheses*.

7.11 ESTIMATION

Estimates of population values, known as *parameters*, are made on the basis of sample values, known as *sample statistics*. Problems in estimation may be divided into two groups.

Point Estimation

Point estimation is the process by which an exact estimate is made of a parameter or population value, that is, a specific number is determined as the estimate of the parameter. This number is derived from a single sample of n observations.

EXAMPLE: A sample mean ($\bar{x}$) is often used as the best estimate of a population mean (μ), and a sample standard deviation (s) is used as the best estimate of a population standard deviation (σ).

Interval Estimation

Interval estimation is the process by which a lower limit and an upper limit are computed within which a parameter is expected to be contained. The interval is known as a *confidence interval.*

EXAMPLE: A statement may be made, such as, for example, "the true arithmetic mean of income recipients in the United States for a particular year is between $4,300 and $4,500." This interval either does or does not contain the true mean, and it is not known whether it does or does not.

Nevertheless, *interval estimation* permits the statistician to attach a probability value to a statement of this type. If he assigns a probability estimate of 0.95, it means that in the long run the statistician will be right 95 per cent of the time in this particular problem.

7.12 THE CONFIDENCE INTERVAL AND THE NORMAL CURVE

Previously *confidence intervals* were set in relation to the normal curve, such that

$\mu \pm 1\sigma$ would contain 68.27 per cent of the items.
$\mu \pm 2\sigma$ would contain 95.45 per cent of the items.
$\mu \pm 3\sigma$ would contain 99.73 per cent of the items.

Often, however, the statistician prefers to use rounded percentages rather than fractional percentages. For example, a 95 per cent confidence interval is more widely used than the 95.45 per cent interval. To drop 0.45 per cent from 95.45 per cent results in a lower z value. Therefore:

$\mu \pm 1.96\sigma$ would contain 95.0 per cent of the items.

A similar adjustment may be made for 99.73 per cent, such that:

$\mu \pm 2.576\sigma$ would contain 99.0 per cent of the items.

Another value which is sometimes used is:

$\mu \pm 1.645\sigma$, which would contain 90.0 per cent of the items.

Estimating the Mean of a Population Using a Large Random Sample

EXAMPLE: Assume that a sample of 100 items is randomly selected from an infinite population, that is, one which is very large. The mean and standard deviation of the sample are computed, such that $\bar{x} = 500$ and $s = 50$. The true mean (μ) is not known. If $\bar{x}$ is the point estimate of the true mean, what is the 95 per cent confidence interval for the population mean?

Solution: The standard error of the mean, or the standard deviation of the probability distribution of sample means, may be estimated as follows:

$$\sigma_{\bar{x}} = \frac{s}{\sqrt{n}}$$

$$= \frac{50}{\sqrt{100}}$$

$$= \frac{50}{10}$$

$$= 5$$

The 95 per cent confidence interval for the true mean is

$$\bar{x} \pm 1.96\ \sigma_{\bar{x}}$$

$$\text{or} \quad 500 \pm (1.96)(5)$$

$$\text{or} \quad 500 \pm 9.80$$

$$500 + 9.8 = 509.8$$

$$500 - 9.8 = 490.2$$

Thus, the range of the 95 per cent confidence interval is from 490.2 to 509.8. This statement means that if repeated random samples were drawn from the same population, and the conclusion each time was that the true mean is within specified limits, the researcher should expect to be right 95 per cent of the time, or in 95 out of 100 tries.

Estimating the Mean of a Population Using a Small Random Sample—the *t* Distribution

THE CONCEPT: In estimating the mean of a population from a large sample, it was assumed that the means of repeated, random samples drawn from the same population would be normally distributed. However, when the sample size is less than 30, the assumption of normality is not valid. The means of small samples distribute themselves in accordance with the *t distribution.*

The normal probability distribution depends upon its mean and standard deviation; the binomial probability distribution depends upon the probability of success in a single event, the number of independent events in a trial, and the stated number of successes; but the *t distribution* depends upon degrees of freedom. A different *t* distribution is available for each number of degrees of freedom.

The smaller the sample size, the more widely dispersed is the *t* distribution. However, as the sample size increases, the *t* distribution blends into the normal probability distribution.

BACKGROUND: The *t distribution* was developed by William S. Gosset (1876–1937), and published in 1908 under the pen name "Student." Employed by the Guinness brewery in Dublin at the time, he published his work under a pen name because of a company rule forbidding such independent work. The distribution is now known as the *Student t distribution.* Gosset proved that the normal distribution could not be applied to a small sample, or one in which the number of items, *n*, is less than 30.

HOW TO READ THE *t* TABLE: Because it is impractical to provide a complete table of areas for all the distributions corresponding to the various sample sizes, a summary of basic information is given in the *t* table, Appendix Table V.

The first column of Table V is headed Degrees of Freedom. This is derived from the sample size, *n*, less 1. Thus, the number of degrees of freedom equals $n - 1$. Each row in Table V provides *t*-distribution information corresponding to a specified number of degrees of freedom.

The column headings are probabilities. The confidence interval corresponding to each probability is as follows:

Probability	*Confidence Interval*
0.10	90%
0.05	95%
0.01	99%

A value from the table under a particular column heading indicates the number of standard error units which must be included on both sides of the sample mean to compute the confidence interval.

INTERPRETATION: When the number of degrees of freedom is small, such as 4 (sample size 5, or $5 - 1 = 4$), then the value for probability .05 is 2.776 instead of the z value 1.960 for the same probability. This means that only 5 per cent of the means of samples of 5 items each are more than 2.776 standard errors above or below the population mean.

The value for probability .01 is 4.604 instead of the z value 2.576 for the same probability. This means that only 1 per cent of the means of samples of 5 items each are more than 4.604 standard errors above or below the population mean.

As the sample size increases, t values in the .05 column approach 1.96, the z value for the 95 per cent confidence interval, and the t values in the .01 column approach 2.576, the z value for the 99 per cent confidence interval. Thus, as sample size increases, the t distribution takes on the shape of the normal curve.

The use of the t table is illustrated below.

EXAMPLE: Five ball bearings are selected at random from the production process. The diameter of each ball bearing is measured in order to estimate the average diameter of ball bearings for the entire production run. The measurements of the five items in the sample yield a mean of .229 inches and a standard deviation of .0082 inches. If $\bar{x}$ is the point estimate of the true mean, what is the 95 per cent confidence interval for the population mean?

Solution: First, compute the standard error of the mean:

$$\sigma_{\bar{x}} = \frac{s}{\sqrt{n-1}}$$

$$= \frac{.0082}{\sqrt{5-1}}$$

$$= \frac{.0082}{\sqrt{4}}$$

$$= \frac{.0082}{2}$$

$$= .0041 \text{ inches}$$

The 95 per cent confidence interval for the true mean, for 4 degrees of freedom ($n - 1 = 5 - 1 = 4$) is obtained from Appendix Table V. For 4 degrees of freedom, and $p = .05$, $t = 2.776$. Therefore,

$$\text{confidence limits} = \bar{x} \pm t_{n-1}\,\sigma_{\bar{x}}$$
$$= .229 \pm (2.776)(.0041)$$
$$= .229 \pm .0113816 \text{ or } .0114$$
$$\text{lower limit} = .229 - .0114 = .2176 \text{ inches}$$
$$\text{upper limit} = .229 + .0114 = .2404 \text{ inches}$$

Thus, the probability is 95 per cent certain that the average diameter of ball bearings in this production run, that is, the true mean, would be between .2176 and .2404 inches.

7.13 TESTING HYPOTHESES

INTRODUCTION: In the sciences and in the social sciences, as well as in everyday life, many important decisions are based very often on judgment alone. In basing a decision on judgment, it is not possible to evaluate mathematically the risk of a wrong decision.

However, if the statistical approach to a problem is feasible, alternative courses of action can be evaluated by the use of statistical decision rules. Mathematical quantification of alternative courses of action is one of the most important contributions of statistical theory and methods. The procedures developed for testing hypotheses provide criteria which, when applicable, help to minimize the probability of making a wrong decision.

The following concepts are basic to statistical decision-making: a *statistical hypothesis*, the *types of error* involved in accepting a false hypothesis or in rejecting a true one, and a *statistical test* for evaluating an hypothesis.

DEFINITION: A *statistical hypothesis* is a theory or tentative statement which is subject to verification. It is based upon observed random variables from which the nature of the true population is estimated. Thus, on the basis of incomplete information, one of two possible decisions is made: Either nature is in state A, or it is not.

To illustrate the meaning of a statistical hypothesis, several examples follow involving two alternatives: Production process A is more economical than production process B, sell or do not sell a particular stock, continue or discontinue production of a given commodity, use treatment

A or treatment B for a particular disease. Or, more simply, drug X is nontoxic or it is toxic.

An *hypothesis* is generally stated in two parts, the *null hypothesis* and the *alternative hypothesis*.

DEFINITION: The *null hypothesis* is the hypothesis to be tested. It is stated negatively, and is therefore called an hypothesis of no difference. The symbol for the null hypothesis is H_o.

EXAMPLE: $H_o \rightarrow$ Drug X is nontoxic.

DEFINITION: The *alternative hypothesis* is the hypothesis against which the null hypothesis is to be tested. It is stated positively, and is called an hypothesis of difference. The symbol for the alternative hypothesis is H_a.

EXAMPLE: $H_a \rightarrow$ Drug X is toxic.

Type I and Type II Errors

Two types of errors may be made. If the null hypothesis is actually true and it is called false, a *Type I error* has been made. If it is actually false, and it is called true, that is, it is not rejected, a *Type II error* has been made.

Thus, a *Type I error* occurs if the null hypothesis is rejected when it is true.

A *Type II error* occurs if the null hypothesis is not rejected when it is false, when actually the alternative hypothesis is true.

These possibilities for error are summarized in Table 7.5.

TABLE 7.5 TYPES OF ERRORS THAT MAY BE MADE IN HYPOTHESIS TESTING

Alternative	Accept H_o	Accept H_a
H_o is true	No error	Type I error
H_a is true	Type II error	No error

Consequences of Type I and Type II Errors

The null and alternative hypotheses are specified prior to initiating the experiment. The statistician seeks to control the risk of a Type I error at 5 per cent. Subject to the first rule, the statistician thus makes the risk of a Type II error as small as possible.

In the example given above, if drug X is nontoxic, then making a Type I error would not lead to serious consequences for the patients who are receiving the drug. If, on the other hand, drug X is toxic, the consequences of accepting the null hypothesis when it is false, a Type II error, would have serious consequences.

A Test of Significance

A statistical test for an hypothesis is one which provides a guide for rejecting or not rejecting an hypothesis. The test is known as a *test of significance.*

DEFINITION: A *test of significance* is one which helps the statistician decide whether a difference between a sample statistic and a population parameter, or between two sample statistics, may be attributed to chance or not.

INTERPRETATION: If the difference is so large that it cannot be ascribed to chance, then the difference is *significant*. This means that differences actually exist. The null hypothesis, which states that no difference exists, is rejected, and the alternative hypothesis is accepted.

If the difference is so small that it may be attributed to chance, it is not *statistically significant*, i.e., it is *insignificant*. The differences may be attributed simply to sample variation, and the null hypothesis is not rejected.

The *significance level* of the test is the probability of committing a Type I error. The probability of making a Type I error is represented by the Greek letter *alpha*, written α. For typical statistical problems, the significance level is set at $p = .05$, or 5 per cent. This is equivalent to a z value of 1.96. In testing hypotheses, the probability of making an *alpha*, or Type I, error may be reduced in two ways: (1) by increasing the size of the sample or (2) by increasing the probability of a Type II error.

The probability of making a Type II error is represented by the Greek letter *beta*, written β. In testing hypotheses, the probability of making a *beta* error may be reduced in two ways: (1) by increasing the size of the sample or (2) by using a larger value for *alpha*. For example, an *alpha* value of 5 per cent ($p = .05$) is larger than an *alpha* value of 1 per cent ($p = .01$).

7.14 GENERAL TYPES OF PROBLEMS IN TESTING HYPOTHESES

Two general types of problem may be noted: (1) problems concerned with the difference between an observed sample statistic and a population parameter which may be either known or hypothetical; (2) problems concerned with the difference between two sample statistics.

Difference Between a Sample Statistic and a Population Value

A random sample of 40 pieces of wire is selected from the output of a firm producing steel wire. The diameters of the 40 pieces of wire are measured, a frequency distribution is formed, and two sample measures are computed, $\bar{x}$, which equals .173 inches, and s, which equals .0119 inches. Are these results consistent with the hypothesis that the average diameter of this firm's output of steel wire (the population) is 0.18 inches? Use the 5 per cent level of significance.

THE LOGIC OF THE PROBLEM: It is assumed that the mean of the population, 0.18 inches, is a known quantity based upon previous experience with this production run. With this information, the problem evaluates the difference between a *sample statistic* and a *population value*, or the *significance of a mean.* The problem seeks to determine whether the mean of a sample ($\bar{x}$) selected at random from a given population is, or is not, truly representative of the parent population.

The method is generally similar to that used in the previous chapter for determining the degree of confidence that can be placed in a given interval containing the true mean. The formula used then was

$$z = \frac{X - \mu}{\sigma}$$

In this problem, however, a decision must be based upon a sampling distribution. An inference concerning a population based upon a sample requires use of the standard error of the mean. Therefore, the formula above is adjusted as follow:

$$z = \frac{\text{mean of the sample} - \text{mean of the population}}{\text{standard error of the mean}}$$

$$= \frac{\bar{x} - \mu}{\frac{s}{\sqrt{n}}}$$

STEPS INVOLVED IN MAKING DECISIONS BASED UPON z:

1. State the null and alternative hypotheses.

2. Set the level of significance. Generally, this is set at a value of 5 per cent, that is, a z value of 1.96.

3. Compute the z value.

4. Make the statistical decision. That is, determine whether the z value is greater than 1.96 or smaller. If it is greater, reject the null hypothesis, and accept the alternative hypothesis. If it is smaller, accept the null hypothesis.

Solution:

1. The null hypothesis (H_o): $\mu = .18$, or no significant difference exists between the mean of the sample ($\bar{x}$) and the mean of the population (μ).

The alternative hypothesis (H_a): $\mu \neq .18$, or a significant difference exists between $\bar{x}$ and μ.

2. The level of significance given in the example is 5 per cent. This is equivalent to a z value of 1.96.

3. Computation of the z value. (If $\bar{x}$ and s are not given, they must be computed from the data contained in the sample.)

$$z = \frac{\bar{x} - \mu}{\frac{s}{\sqrt{n}}}$$

$$= \frac{.173 - .18}{\frac{.0119}{\sqrt{40}}}$$

$$= \frac{-.007}{\frac{.0119}{6.32}}$$

$$= \frac{-.007}{.0019}$$

$$= -3.684$$

4. The statistical decision. The decision rule in this problem is: Reject H_o if z is less than -1.96 or greater than $+1.96$. Since 3.684 is greater than 1.96, the null hypothesis (H_o) is rejected, and the alternative hypothesis (H_a) is accepted, that is, the population mean does not equal .18.

Based upon the sample, a *confidence interval estimate of the population mean* diameter with a 95 per cent level of confidence is computed as follows:

$$\text{confidence interval} = \bar{x} \pm (1.96)(\sigma_{\bar{x}}) \qquad \text{where } \sigma_{\bar{x}} = \frac{s}{\sqrt{n}}$$

$$= .173 \pm (1.96)\left(\frac{.0119}{\sqrt{40}}\right)$$

$$= .173 \pm (1.96)\left(\frac{.0119}{6.32}\right)$$

$$= .173 \pm (1.96)(.0019)$$

$$= .173 \pm .0037$$

$$\text{upper limit} = .173 + .0037 = .1767 \text{ inches}$$

$$\text{lower limit} = .173 - .0037 = .1693 \text{ inches}$$

$$\text{confidence interval} = .1693 \text{ inches to } .1767 \text{ inches}$$

Thus, 95 out of a 100 samples drawn from this population would yield an average mean diameter between .1693 and .1767 inches. An hypothesis with a mean population of less than .1693 inches or more than .1767 inches would be rejected.

Difference Between Two Sample Statistics

Two brands of tires are subjected to standard wear tests to determine which brand has a longer life. A sample of 50 of brand A tires and 50 of brand B, a competitor's product, are selected and tested. The mean and standard deviation for the sample of each brand are indicated below. Is the difference between the average life of these two brands of tires significant at the 5 per cent level?

Brand A	*Brand B*
$n_1 = 50$	$n_2 = 50$
$\bar{x}_1 = 42$ months	$\bar{x}_2 = 45$ months
$s_1 = 12$ months	$s_2 = 16$ months

THE LOGIC OF THE PROBLEM: In this problem the purpose is to find out whether the two samples have been drawn from the same population or from two different populations. It illustrates an application which is similar to the problem of comparing the sample mean of a "control" group with the sample mean of an "experimental" group. The z formula is used. In this case:

$$z = \frac{\text{the difference between the means of the first sample } (\bar{x}_1) \text{ and the second sample } (\bar{x}_2)}{\text{the standard error of the difference of the means } (\sigma_{\bar{x}_1-\bar{x}_2})}$$

where

$$\sigma_{\bar{x}_1-\bar{x}_2} = \sqrt{\frac{s_1^2}{n_1} + \frac{s_2^2}{n_2}}$$

This is based upon the mathematically proved theorem which states that the standard error of the difference between two independent random variables is equal to the square root of the sum of their variances. Actually,

$$z = \frac{(\bar{x}_1 - \bar{x}_2) - (\mu_1 - \mu_2)}{\sqrt{\frac{s_1^2}{n_1} + \frac{s_2^2}{n_2}}}$$

According to H_o (the null hypothesis), $\mu_1 = \mu_2$. Therefore, the difference $(\mu_1 - \mu_2)$ equals 0, and the z formula becomes

$$z = \frac{\bar{x}_1 - \bar{x}_2}{\sqrt{\frac{s_1^2}{n_1} + \frac{s_2^2}{n_2}}}$$

The steps involved in making a decision based upon z are similar to the example illustrating the difference between a sample statistic and a population value. Only the formula is different.

Solution:

1. The null hypothesis (H_o): No significant difference exists between the average life of tires for brand A and brand B. The alternative hypothesis (H_a): A significant difference exists between the average life of tires for brand A and brand B.

2. The level of significance is given in the example as 5 per cent. This is equivalent to a z value of 1.96.

3. Computation of the z value:

$$
\begin{aligned}
z &= \frac{\bar{x}_1 - \bar{x}_2}{\sqrt{\dfrac{s_1^2}{n_1} + \dfrac{s_2^2}{n_2}}} \\
&= \frac{42 - 45}{\sqrt{\dfrac{144}{50} + \dfrac{256}{50}}} \\
&= \frac{3}{\sqrt{2.88 + 5.12}} \\
&= \frac{3}{\sqrt{8.00}} \\
&= \frac{3}{2.83} \\
&= 1.06
\end{aligned}
$$

4. The answer, 1.06, is less than 1.96. Therefore, accept the null hypothesis, i.e., that no difference exists between the average life of tire wear for brand A and brand B. The difference between $\bar{x}_1$ and $\bar{x}_2$ is due merely to chance, and the quality of the brands is almost equivalent.

7.15 DETERMINATION OF SAMPLE SIZE

The size of the sample which is required in order to make a valid inference about a population is dependent upon *several factors.* The factors which must be considered are:

1. The degree of reliability desired
2. The amount of time and money available for the study or survey
3. The amount of information available about the population from which the sample is to be selected

Degree of Reliability Desired

Reliability is determined by the confidence interval derived from the standard error of the mean. It indicates the extent to which a similar result may be expected if a second or a third sample of n items is selected from the same population. The larger the sample size, the greater the reliability which may be expected. This may be illustrated by the formula for the standard error of the mean.

$$\sigma_{\bar{x}} = \frac{\sigma}{\sqrt{n}}$$

The width of the confidence interval is determined by the standard error of the mean, which, in turn, is directly dependent upon the sample size. As the sample size, n, is increased, the standard error of the mean decreases; the confidence interval is narrower; and the inferences made about the population will be more reliable. A sample size of 2,000 items will give more consistently reliable estimates than a sample of 100 items.

Time and Money Available for the Study or Survey

A sample size of 2,000 items will cost more and take longer than a sample of only 100 items. Costs, therefore, will vary directly with the sample size. The researcher must be aware of available resources against which to evaluate the degree of accuracy and reliability desired.

Information About the Population

In the formula used for determining sample size, a prior estimate of the standard deviation of the population (σ) is required. This estimate may be based upon previous studies or surveys of similar or related data.

Applications

FORMULA: The formula for the sample size is

$$n = \frac{z^2\sigma^2}{(\bar{x} - \mu)^2}$$

where z = z value required at a specified level of confidence
σ = standard deviation of the population
$\bar{x}$ = sample mean
μ = population mean
n = sample size

The value of μ may be hypothetical. Only the desired difference between $\bar{x}$ and μ is predetermined.

DERIVATION OF THE FORMULA: The formula for the sample size is derived from the basic z formula

$$z = \frac{\bar{x} - \mu}{\frac{\sigma}{\sqrt{n}}}$$

$$z \frac{\sigma}{\sqrt{n}} = \bar{x} - \mu$$

$$\frac{z\sigma}{(\bar{x} - \mu)} = \sqrt{n}$$

$$n = \frac{z^2\sigma^2}{(\bar{x} - \mu)^2}$$

EXAMPLE: A market analyst has information from previous sample surveys that the population standard deviation is 16. He wishes to determine the size of a sample such that a difference of 3 between a sample mean and a population mean will be significant at the 5 per cent level. What sample size is required?

Solution:

$$\begin{aligned} n &= \frac{z^2\sigma^2}{(\bar{x} - \mu)^2} \\ &= \frac{(1.96)^2(16)^2}{3^2} \\ &= \frac{(3.8416)(256)}{9} \\ &= \frac{983.4496}{9} \\ &= 109 \end{aligned}$$

A sample size of about 110 would yield the desired results.

The problem of designing a sample and determining the sample size is usually referred to a highly skilled statistical technician who has had advanced training in probability statistics. However, for specified finite population sizes, the number of sample units required to provide valid conclusions at various confidence intervals has been computed. These sample sizes have been derived by modifying the sample size formula used above in order to reflect the size of the finite population to be sampled.

The figures in Table 7.6 indicate the number of sample units required for specified finite populations in order to obtain a 95 per cent confidence interval. Even though Table 7.6 is based upon finite populations of occupied housing units in project areas, it is equally applicable to other finite populations which are to be sampled. Using the number of sample units indicated will provide a 95 per cent confidence interval.

TABLE 7.6 SAMPLE SIZE REQUIRED FOR FINITE POPULATIONS FOR 95 PER CENT CONFIDENCE INTERVAL
(number)

Occupied housing units in project area	Sample units required
500	222
1,000	286
1,500	316
2,000	333
2,500	345
3,000	353
3,500	359
4,000	364
4,500	367
5,000	370
6,000	375
7,000	378
8,000	381
9,000	383
10,000	385
15,000	390
20,000	392
25,000	394
50,000	397
100,000	398
∞	400

SOURCE: Herbert Arkin and Raymond R. Colton, *Tables for Statisticians*, New York, Barnes and Noble, 1950, Table 19, p. 136. Reproduced from *A Method for Employing Sampling Techniques in Housing Surveys*, New York State Division of Housing, September 1948.

7.16 THE CHI-SQUARE TEST OF GOODNESS OF FIT

THE CONCEPT: While the normal, binomial, and Student t distributions are useful in solving a variety of problems, some problems require other probability distributions. The chi-square probability distribution is one of these.

One of the many important applications of the chi-square distribution is in determining whether differences between observed frequencies and theoretical or expected frequencies may be attributed to chance. If the theoretical frequencies are normally distributed, it is possible to judge such a relationship by superimposing an observed frequency distribution on a normal curve, as in Figure 7.3. In the figure, the shaded areas represent disagreement between observed and theoretical frequencies. The chi-square test of goodness of fit evaluates this relationship mathematically by assigning a numerical value to the shaded areas.

BACKGROUND: *The chi-square test of goodness of fit* compares theoretical frequencies, also called expected or computed frequencies, with observed frequencies in order to ascertain if the discrepancy is or is not greater than might be expected to occur by chance.

FORMULA:

$$\chi^2 = \Sigma\left(\frac{(f_o - f_t)^2}{f_t}\right)$$

where Σ = sum of
f_o = observed or actual frequencies
f_t = theoretical (expected or computed) frequencies
χ^2 = chi-square

INTERPRETATION OF THE VALUE OF χ^2: The curve fit is good if the χ^2 value is small, and bad if the χ^2 value is large. If $\chi^2 = 0$, agreement between observed and theoretical frequencies is perfect. The greater the discrepancy between observation and expectation, the larger will be χ^2.

HOW TO READ THE χ^2 TABLE: Table VI, page 282, shows values for the χ^2 distribution. The first column, n, is the number of degrees of freedom, derived from $n - 1$, where n represents the number of classes or items. The column heads are probabilities, indicating the probability for χ^2 to be greater than the value indicated in the table. If the computed value for χ^2 is greater than appears in the table for some specified probability, this is interpreted to indicate that the differences between observed and theoretical frequencies are too large to be attributed solely to chance. The null hypothesis is rejected, and the alternative hypothesis is accepted.

EXAMPLE 1: The data in Table 7.7 indicate the length of life of 1,200 light bulbs selected at random from a production run. Also indicated are the theoretical frequencies or expected normal curve frequencies which could have been expected if the data were actually a normal distribution with the same mean and standard deviation as the given data. Does the normal curve provide a good fit for the observed data?

Solution:

1. State the null and alternative hypotheses: The null hypothesis (H_o): No difference exists between the observed and theoretical frequencies. The alternative hypothesis (H_a): Significant differences exist between observed and theoretical frequencies.

2. Set up a table for deriving value of chi-square similar to Table 7.7.

TABLE 7.7 CHI-SQUARE TEST OF AVERAGE LIFE SPAN OF 1,200 LIGHT BULBS

Life in hours (l_1–l_2)	Number of bulbs: Observed frequencies (f_o)	Theoretical frequencies (f_t)	($f_o - f_t$)	$(f_o - f_t)^2$	$\frac{(f_o - f_t)^2}{f_t}$
10–14.99	8	6	2	4	0.67
15–19.99	23	31	− 8	64	2.06
20–24.99	97	110	−13	169	1.53
25–29.99	224	235	−11	121	0.51
30–34.99	417	326	91	8,281	25.40
35–39.99	242	279	−37	1,369	4.90
40–44.99	112	150	−38	1,444	9.08
45–49.99	57	51	6	36	0.70
50–54.99	16 } 20[a]	11 } 12[a]	8	64	5.33
55–59.99	4	1			
Total	1,200	1,200			$\chi^2 = 50.18$

[a] The last two classes of frequencies were combined to fulfill the requirement that each class contain at least 5 items.

3. This problem contains 10 classes. Therefore, $n = 10$. Degrees of freedom for this problem: $n - 3 = 7$. Three is deducted, because three values were used in the computation of the theoretical frequencies: the

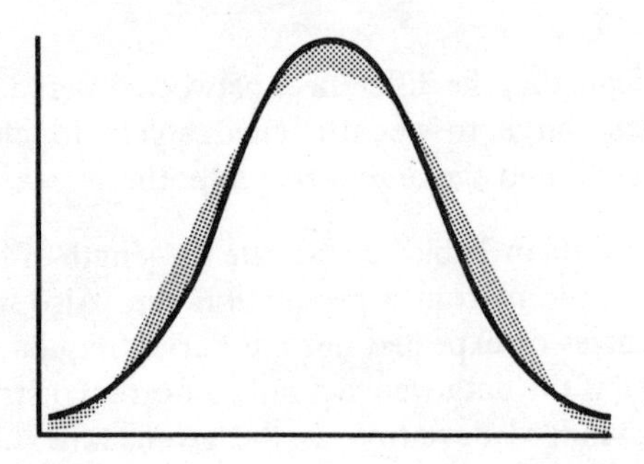

Fig. 7.3 An Observed Frequency Distribution Superimposed on a Normal Curve

total frequency, the mean, and the standard deviation. While theoretical frequencies were given in this problem, when not given they must be computed. From Table VI, page 282, χ^2 with 7 d.f. at $p = .05$, equals 14.067. The computed value of $\chi^2 = 50.18$. The normal curve provides a poor fit: $50.18 > 14.067$. Therefore, the null hypothesis is rejected, and the alternative hypothesis is accepted.

EXAMPLE 2 illustrates a second very important application of the chi-square distribution. The data in Table 7.8 summarize the results of the votes of 217 students who were requested to designate their preference

TABLE 7.8 OBSERVED STUDENT PREFERENCES FOR TWO BRANDS OF SOAP

Soap	Student preference		
	For	Against	Total
Brand A	19	79	98
Brand B	49	70	119
Total	68	149	217

for or against two brands of soap. Do these results depart significantly from the proportions of votes *for* and *against* each brand that might be expected if it were assumed that one brand was not preferred over the other? The data in Table 7.8 are observed frequencies recording actual votes of students for and against each brand of soap. The format of the table is known as a 2-by-2 contingency table.

DEFINITION: A 2×2 *contingency table* is the general term for a two-way classification specifying varying numbers of discrete categories in each of two dimensions.

Solution:

1. State the null and alternative hypotheses: The null hypothesis (H_o): No difference is discernible in the preference for either one of the two brands of soap. The alternative hypothesis (H_a): A significant difference exists in the preference for one or the other of the two brands of soap.

2. Compute the theoretical frequencies: If no preference existed for either one of the two brands, then the proportions for and against each would be in exactly the same proportion as the total votes cast for each brand. Thus, the percentages would be as follows: 98 ÷ 217 = 45.2 per cent; 119 ÷ 217 = 54.8 per cent. The expected or theoretical frequencies for brand A are 45.2 per cent × 68 = 31, and 45.2 per cent × 149 = 67. The expected or theoretical frequencies for brand B are 54.8 per cent × 68 = 37, and 54.8 per cent × 149 = 82. These theoretical values are indicated in Table 7.9.

TABLE 7.9 THEORETICAL STUDENT PREFERENCES FOR TWO BRANDS OF SOAP

Soap	Theoretical votes For	Against	Total
Brand A	31	67	98
Brand B	37	82	119
Total	68	149	217

3. The computation of chi-square (χ^2) is carried out in Table 7.10.

TABLE 7.10 COMPUTATION OF CHI-SQUARE FOR A TWO-BY-TWO CONTINGENCY TABLE OF STUDENT PREFERENCES FOR TWO BRANDS OF SOAP

Frequencies Observed (f_o)	Theoretical (f_t)	$(f_o - f_t)$	$(f_o - f_t)^2$	$\frac{(f_o - f_t)^2}{f_t}$
19	31	−12	144	4.6
49	37	12	144	3.9
79	67	12	144	2.1
70	82	−12	144	1.8
				$\chi^2 = 12.4$

4. Degrees of freedom: Degrees of freedom in this problem are 4 − 3 or 1, because as soon as one entry is made in a 2-by-2 contingency table, the other three entries are derived by subtraction from the row and column totals. Hence, only one degree of freedom is available. For example, as soon as 31 is entered in the first cell, as follows:

31		98
		119
68	149	217

the only values possible for the other three cells are 67, 37, and 82.

31	67	98
37	82	119
68	149	217

5. Level of significance: If a 5 per cent level of significance is selected ($p = .05$), then from the table of chi-square with d.f. = 1, and $p = .05$, the χ^2 value, from Table VI, is 3.841.

6. Make decision: The rules for not rejecting (accepting) or rejecting of the H_o are as follows:

If the computed value of $\chi^2 >$ the table value, reject H_o, and accept H_a.
If the computed value of $\chi^2 \leq 3.841$, do not reject (accept) H_o.

In this problem, computed $\chi^2 = 12.4$, which is greater than 3.841. The null hypothesis is rejected, and the alternative hypothesis is accepted: A significant difference exists in preference for one of the two brands of soap. Apparently, brand B is preferred over brand A.

7.17 OTHER DISTRIBUTIONS

Other probability distributions have been identified. The *F distribution*, listed as the fourth of the continuous probability distributions in Chapter 6, is outside the scope of this book. Applications of the *F* distribution are included in the branch of statistics known as *analysis of variance*.

EXERCISES

1. Define or explain the following:

Alternative hypothesis
Area sampling
Central limit theorem
Chi-square distribution
Chi-square test
Cluster sampling
Confidence interval
Degrees of freedom
Difference between two sample means
Double sampling
Estimation
Heterogeneous data
Homogeneous data
Interval estimation
Judgment sample
Large sample
Null hypothesis
Parameter
Point estimation
Probability sample
Purposive sample
Quota sample
Random sample
Rectangular distribution
Sample
Sample coverage
Sampling distribution of the arithmetic mean
Sequential sample
Significance level
Significance of a mean
Small sample
Standard error of the mean
Statistical hypothesis
Statistical inference
Stratified sample
Stratum
Systematic sample
t distribution
Table of random digits
Testing hypotheses
Test of significance
Two-by-two contingency table
Type I error
Type II error
Universal coverage

2. Using the data in Table 7.11, do the following:

   a. Draw a random sample of 75 items using Table IV, Random Digits, page 280.
   b. Prepare a frequency distribution.
   c. Compute the arithmetic mean and standard deviation.

3. Given a population with a true mean of 500 and a standard deviation of 50,

   a. What is the standard error of the mean of a sample of 75 selected from this population?

TABLE 7.11 GROSS MONTHLY RENTS OF 500 RENT-CONTROLLED APARTMENTS IN NEW YORK CITY, 1964

001	$107.50	051	$92.00	101	$68.00	151	$44.00	201	$80.00
002	57.00	052	135.00	102	44.45	152	75.00	202	58.00
003	115.00	053	95.00	103	38.20	153	129.00	203	36.40
004	59.44	054	56.00	104	76.00	154	37.95	204	64.00
005	32.35	055	60.00	105	75.00	155	72.00	205	38.95
006	74.75	056	75.00	106	54.13	156	94.60	206	38.57
007	50.00	057	51.11	107	32.00	157	33.00	207	43.60
008	60.84	058	32.65	108	26.45	158	75.00	208	25.11
009	72.25	059	95.00	109	53.47	159	77.50	209	36.50
010	32.25	060	35.00	110	25.00	160	58.00	210	36.95
011	85.00	061	104.00	111	158.00	161	43.00	211	120.00
012	86.00	062	52.09	112	60.00	162	113.55	212	26.80
013	100.00	063	56.00	113	86.66	163	80.00	213	29.84
014	39.75	064	62.05	114	46.81	164	88.66	214	64.50
015	56.20	065	80.00	115	52.04	165	45.00	215	27.25
016	49.50	066	86.00	116	65.00	166	78.40	216	90.00
017	75.00	067	65.00	117	33.10	167	86.00	217	30.00
018	42.00	068	75.45	118	49.75	168	68.00	218	43.05
019	41.55	069	109.50	119	50.50	169	100.00	219	58.00
020	48.00	070	106.00	120	27.00	170	66.00	220	43.60
021	40.42	071	61.99	121	55.00	171	90.00	221	36.40
022	25.00	072	80.00	122	62.20	172	49.98	222	100.00
023	98.35	073	28.75	123	66.00	173	63.00	223	42.32
024	52.45	074	40.00	124	56.00	174	62.92	224	59.66
025	91.25	075	71.50	125	52.00	175	129.00	225	74.00
026	69.00	076	35.00	126	26.00	176	34.39	226	40.00
027	54.75	077	43.19	127	64.40	177	78.40	227	20.00
028	30.35	078	55.50	128	77.40	178	37.39	228	64.50
029	52.00	079	145.00	129	26.00	179	57.00	229	61.50
030	85.00	080	71.00	130	74.75	180	65.00	230	90.00
031	75.00	081	60.00	131	107.50	181	44.10	231	69.00
032	75.00	082	72.22	132	40.00	182	30.00	232	70.00
033	64.00	083	100.00	133	60.00	183	75.00	233	54.20
034	58.00	084	57.67	134	65.00	184	25.00	234	62.30
035	78.00	085	28.30	135	65.39	185	80.00	235	57.50
036	179.50	086	28.00	136	35.00	186	80.00	236	89.93
037	158.00	087	73.93	137	65.00	187	75.75	237	100.00
038	96.75	088	53.47	138	80.00	188	77.44	238	103.44
039	115.00	089	81.70	139	117.00	189	60.00	239	94.60
040	100.00	090	41.07	140	59.00	190	103.35	240	25.00
041	73.53	091	46.29	141	65.00	191	73.10	241	37.75
042	47.22	092	22.00	142	66.65	192	50.00	242	55.05
043	24.00	093	72.00	143	77.50	193	22.00	243	38.00
044	140.00	094	107.50	144	80.00	194	24.32	244	75.00
045	115.00	095	51.16	145	72.50	195	40.00	245	81.50
046	179.50	096	78.28	146	35.00	196	68.50	246	85.00
047	124.00	097	50.00	147	33.50	197	51.75	247	49.80
048	100.00	098	60.00	148	69.00	198	100.00	248	29.55
049	43.70	099	90.00	149	44.15	199	86.50	249	64.00
050	115.00	100	54.22	150	45.00	200	25.12	250	23.00

(Table 7.11 continued)

TABLE 7.11 (CONTINUED)

251	$80.00	301	$41.73	351	$39.35	401	$175.00	451	$83.00
252	28.75	302	40.00	352	49.00	402	65.00	452	85.00
253	60.00	303	40.00	353	62.00	403	85.00	453	80.00
254	28.00	304	49.15	354	60.00	404	53.75	454	60.00
255	32.00	305	90.30	355	42.32	405	50.82	455	23.00
256	61.00	306	38.00	356	24.00	406	96.00	456	45.00
257	91.00	307	86.00	357	31.75	407	60.00	457	42.00
258	52.00	308	60.20	358	59.20	408	75.00	458	46.00
259	90.30	309	55.90	359	33.00	409	48.00	459	34.50
260	42.00	310	86.00	360	47.50	410	44.30	460	50.00
261	50.00	311	80.00	361	64.99	411	90.00	461	43.70
262	64.83	312	34.00	362	51.75	412	46.00	462	66.00
263	45.00	313	61.00	363	42.45	413	133.20	463	80.00
264	53.75	314	63.00	364	43.70	414	193.00	464	45.55
265	55.90	315	50.00	365	75.50	415	79.00	465	75.00
266	40.00	316	69.00	366	42.00	416	28.75	466	48.80
267	90.00	317	98.90	367	72.00	417	55.90	467	30.00
268	56.00	318	51.15	368	46.37	418	48.49	468	27.15
269	90.30	319	62.00	369	39.95	419	46.00	469	35.37
270	36.31	320	59.15	370	28.22	420	71.00	470	57.46
271	86.00	321	50.70	371	69.33	421	75.00	471	63.25
272	43.30	322	62.00	372	80.00	422	78.00	472	48.94
273	51.50	323	59.55	373	50.00	423	44.00	473	64.73
274	130.00	324	46.00	374	49.00	424	78.00	474	39.55
275	47.80	325	80.00	375	25.00	425	28.00	475	50.00
276	82.00	326	30.00	376	78.00	426	21.00	476	76.70
277	35.00	327	63.80	377	38.00	427	70.00	477	55.00
278	85.00	328	60.00	378	65.00	428	60.00	478	51.40
279	90.30	329	80.00	379	46.50	429	63.05	479	44.55
280	47.30	330	64.85	380	48.90	430	50.00	480	52.75
281	72.00	331	38.00	381	32.20	431	42.00	481	27.45
282	44.00	332	82.50	382	34.40	432	86.00	482	40.00
283	70.00	333	125.00	383	34.50	433	48.15	483	58.35
284	90.00	334	50.00	384	80.00	434	40.00	484	30.00
285	23.15	335	34.85	385	98.35	435	34.50	485	52.75
286	64.00	336	27.76	386	86.67	436	46.45	486	94.60
287	60.00	337	21.70	387	40.00	437	58.00	487	94.60
288	42.00	338	27.00	388	78.00	438	109.00	488	81.00
289	39.55	339	43.00	389	51.75	439	94.00	489	79.50
290	48.00	340	55.00	390	90.85	440	78.00	490	60.00
291	39.67	341	63.50	391	46.42	441	52.90	491	48.70
292	23.35	342	28.00	392	36.45	442	62.00	492	43.50
293	66.00	343	56.00	393	42.62	443	100.00	493	64.00
294	27.25	344	57.50	394	79.00	444	50.00	494	60.00
295	22.48	345	40.00	395	63.00	445	31.00	495	70.00
296	135.00	346	77.40	396	55.50	446	50.00	496	43.00
297	47.30	347	21.13	397	80.00	447	28.75	497	52.75
298	37.00	348	44.00	398	72.40	448	48.30	498	60.00
299	29.50	349	27.00	399	85.00	449	74.75	499	69.00
300	48.00	350	53.50	400	78.00	450	67.32	500	65.00

b. What is the standard error of the mean of a sample of 150 selected from this population?
c. If a large number of samples of 75 items are selected from this population, 68.27 per cent of the sample means should be between ______ and ______.
d. If a large number of samples of 150 items are selected from this population, 95.0 per cent of the sample means should be between ______ and ______.
e. If a sample of 150 items is selected, what is the probability that the sample mean will exceed 510?

4. A random sample of 50 stampings is selected from a firm's output. The thickness of each stamping is measured, a frequency distribution is formed, and two sample measures are computed: $\bar{x} = .165$ inches, and $s = .0129$. Are these results consistent with the hypothesis that the average thickness of stampings produced by this firm (the population) is .15 inches? Use the 5 per cent level of significance.

5. To study the daily unit output of its workers, Plant A seeks to compare its production record with Plant B of the same company. A sample of 50 workers is used in the study, and the following results are obtained:

Plant A	*Plant B*
$n_1 = 50$	$n_2 = 50$
$\bar{x}_1 = 217$ units	$\bar{x}_2 = 230$ units
$s_1 = 71$ units	$s_2 = 48$ units

a. State the null hypothesis and the alternative hypothesis.
b. Compute z.
c. Using a 95 per cent confidence interval, should the null hypothesis be accepted or rejected?
d. Is the mean production of Plants A and B significantly different, or is the difference insignificant, that is, due only to chance factors?

6. An advertising firm is preparing to conduct a sample survey. Previous surveys have indicated the value of the population standard deviation as 24. Determine the size of the sample such that a difference of 5 between a sample mean and a population mean will be significant at the 5 per cent level.

7. Students are requested to state their preference for or against two textbooks. The results of this inquiry are given in Table 7.12.

TABLE 7.12 STUDENT PREFERENCES FOR TWO TEXTBOOKS

	Student preference		
Text	For	Against	Total
Smith	20	80	100
Jones	50	71	121
Total	70	151	221

a. State the null hypothesis and the alternative hypothesis.
b. Set up a work table as shown in Table 7.13.

TABLE 7.13 WORK TABLE FOR EXERCISE 7b

f_o	f_t	$(f_o - f_t)$	$(f_o - f_t)^2$	$\frac{(f_o - f_t)^2}{f_t}$
20	32			
50	38			
80	68			
71	83			

c. Explain the derivation of the theoretical frequencies indicated above.
d. Compute chi-square.
e. Using a 5 per cent level of significance, should the null hypothesis be accepted or rejected? Explain the reasoning.

8

Linear Regression and Correlation

This chapter is concerned with the analysis and description of the relationship between two variables. The study of the degree of association between two variables is called regression analysis *and* correlation analysis.

In regression analysis *the two variables are related in such a way that it is possible to predict the value of one of the variables on the basis of knowledge about the second variable. The relationship between advertising expenditures and the volume of sales is an example of regression analysis.*

In correlation analysis *the degree of association between two variables is also important, but the selection of the variable which is designated as dependent is a matter of preference. The relationship between height and weight of* 10*-year old boys is an example of correlation analysis.*

The methods of regression analysis *and* correlation analysis *are extensions of techniques presented in previous chapters. The calculation of measures of location and measures of variation is based upon observations of a single variable. When these techniques are extended to analysis of the relationships between two variables, the statistician has a useful tool for answering many types of interesting questions.*

This chapter considers first the concept, and then the simplest device, for studying two related variables: the scatter diagram. An example of the scatter diagram is shown.

This is followed by the types of relationship between two variables and an analysis of the basic concepts and measures, including the regression line, the standard error of estimate, the coefficient of correlation, the coefficient of determination, and the coefficient of nondetermination.

A summary of the concept formulas and computational forms as well as the steps in analyzing linear regression and correlation provides the tools and the procedure for solving problems. A detailed example then illustrates the application of the methodology.

8.1 THE RELATIONSHIP BETWEEN TWO VARIABLES

The techniques concerned with the measurement of the relationship between two variables are sometimes referred to as covariation. The simplest form of relationship between two variables is a linear, or straight line, relationship. In some cases, however, the relationship may be curvilinear, or curved line.

In linear, or simple, regression and correlation analysis, the two variables are (1) the X variable, or independent variable, and (2) the Y variable, or dependent variable. When regression analysis is used as a tool in forecasting, the variable X is used to make a forecast about a variable Y. The X variable, therefore, is known as the *predicting variable*, and the Y variable as the *predicted variable.*

Definitions

Linear regression is a statistical technique which measures and analyzes the relationship between two variables, X and Y, which can be approximated by a straight line. Thus, for every value of X, the independent variable, a value for Y may be predicted, based upon the linear equation $Y_c = a + bX$, where Y_c is the computed Y.

Linear correlation is a statistical technique which measures and analyzes the degree to which two variables, X and Y, fluctuate with reference to one another. While the relationship between the X and Y variables may be approximated by a straight line, $Y_c = a + bX$, the linear equation does not imply a cause-effect relationship.

Graphing the Two Variables

On a graph of the two variables, each X value is plotted on the X axis, and each Y value is plotted on the Y axis. Each X value and its related Y value make up the coordinates of a point. After all the points have been plotted on a graph, the result is known as a *scatter diagram*. The scatter diagram helps to visualize the relationship, if any, between the X and Y variables.

Fig. 8.1 Positive Correlation

Fig. 8.2 Negative Correlation

DEFINITION: A *scatter diagram* is a graph showing each pair of observations as a dot.

8.2 TYPES OF RELATIONSHIP

The solutions sought include the following types of relationships:

1. Do the series vary together in the same direction [positive (+) correlation]?
2. Do the series vary together in opposite directions [negative (−) correlation]?
3. Do they not vary together at all?
4. Do the series have a perfect relationship?

Positive (+) Correlation

DEFINITION: Low values of one variable go with low values of the other variable, and high values of one variable go with high values of the other variable.

EXAMPLE: The relationship between advertising expenditures and the volume of sales, depicted in Figure 8.1, is positively correlated.

Negative (−) Correlation

DEFINITION: High values of one variable go with low values of the other variable, and low values of one variable go with high values of the other variable.

EXAMPLE: The relationship between advancing age of older workers and their volume of output, depicted in Figure 8.2, is negatively correlated.

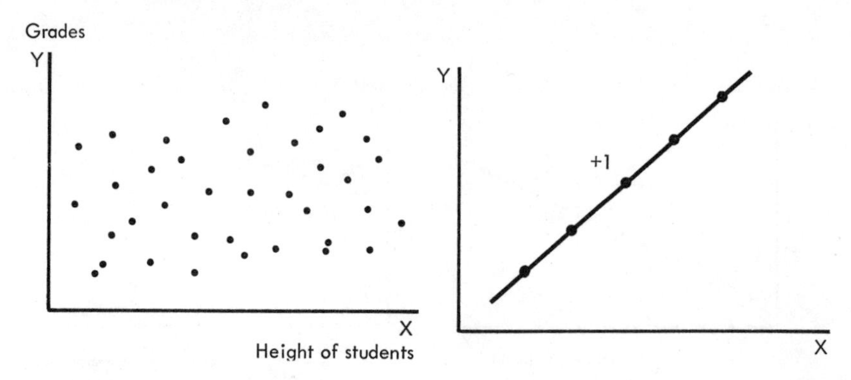

Fig. 8.3 No Correlation **Fig. 8.4** Perfect Positive Correlation

No Correlation

DEFINITION: No relationship exists between the two variables.

EXAMPLE: There is no relationship between height of students and their grades, as depicted in Figure 8.3.

Perfect Correlation

DEFINITION: The points indicating the relationship between the X and Y variables form a straight line rather than merely a path. Perfect correlation can be either positive $(+1)$ or negative (-1).

EXAMPLE: Generally, perfect correlation is never found in the practical, applied problems of business or economics (Figures 8.4 and 8.5).

8.3 THE LEAST SQUARES METHOD

Measures of correlation may be obtained either by the least squares method or by the product moment method. The least squares method is presented here; the product moment method, in Sections 8.12 and 8.13.

The Concept

The *least squares method* is based upon the principle that a line of best fit, or one which describes a relationship between two variables best, is a line for which the sum of the squares of the deviations, or differences between values on the straight line itself and actual values, will be a minimum. Only one line from among the infinite number of lines which may be drawn can meet this requirement. The line of best fit must be computed mathematically.

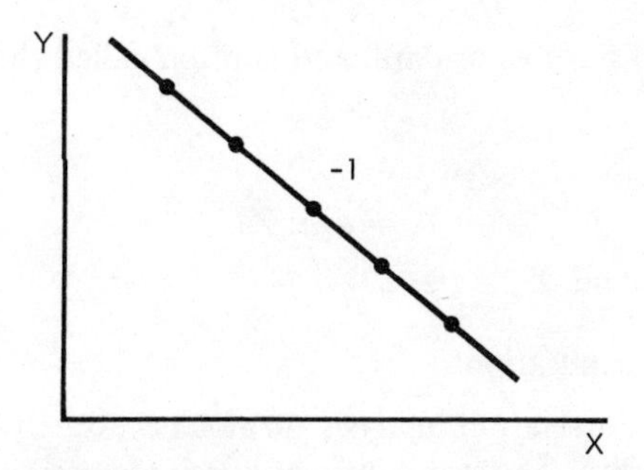

Fig. 8.5 Perfect Negative Correlation

The method of least squares was invented by Gauss in 1795, and represents the beginning of his interest in the theory of errors of observation, which resulted ultimately in the Gaussian law of the normal distribution of errors.

8.4 THE FIVE BASIC CONCEPTS AND MEASURES USED IN LINEAR REGRESSION AND CORRELATION

1. The regression line
2. The standard error of estimate
3. Coefficient of correlation
4. Coefficient of determination
5. Coefficient of nondetermination

8.5 THE REGRESSION LINE

THE CONCEPT: The concept of the regression line was developed by Sir Francis Galton in the middle of the nineteenth century in the course of his studies of heredity. He found that the children of tall parents were not necessarily tall, and that the children of short parents were not necessarily short, but that children tend to regress back toward the average height of the race. Thus height of offspring tend to go back or *regress* toward the average, or the the *line of regression*. This term is still used.

DEFINITION: The *regression line* describes the nature of the relationship between two variables. It is the line of best fit, i.e., the arithmetic mean of the squares of vertical deviations is smaller when computed from this line than from any other. The regression line is the same in concept as the arithmetic mean of a series of data, and is often called the line of average relationship.

FORMULA:

$$Y_c = a + bX$$

where Y_c = computed Y.

Correlation and Causation

Regression and correlation analysis do not necessarily signify a cause-and-effect relationship. The techniques of linear regression and correlation measure the nature and degree of association or covariation between the two variables, but the procedures described do not indicate causality or the lack of causality between two variables. For example, wages in one of the building trades show a high degree of positive correlation with sales in a particular chain of retail stores. This case of correlation does not manifest causation. Both series merely reflect a continuing upswing in the economy, a third variable, which bears a significant impact on both wages and business activity. Thus, the relationship between wages and sales in this illustration is statistical and does not indicate causality.

Spurious Correlation

Further along these lines, it is also appropriate to point out that a statistical relationship between two variables may exist even when no logical basis for the relationship can be ascertained. This is referred to as spurious correlation or nonsense correlation. The classic example of spurious correlation is the close covariation which exists between teachers' salaries and the consumption of liquor. In this illustration two elements have been correlated which in reality bear no relationship to each other.

8.6 THE STANDARD ERROR OF ESTIMATE

THE CONCEPT: When a forecast of Y is made that is based upon X, the standard error of estimate is used to answer the question, how good is the estimate? The standard error of estimate, therefore, is used in the same way as the standard deviation. While the standard deviation measures dispersion around the arithmetic mean, the standard error of estimate is a measure of dispersion around the line of regression, and is a good example of interval estimation.

The more closely the dots cluster around the line of regression, the more representative the line is, and the better the estimates based upon the

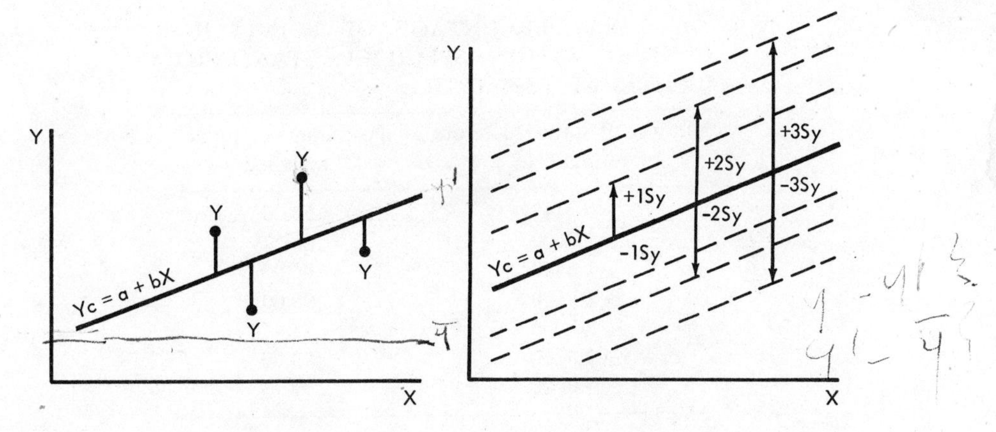

Fig. 8.6 A Line of Regression for Two Variables

Fig. 8.7 Standard Error of Estimate Bands Around Line of Regression

equation for this line. In this illustration, the standard error of estimate would be small.

If the scatter around the regression line is large, the standard error of estimate would be large. The relationship between the X and Y variables, as measured by the line of regression, would be small, and the forecasts made will be less accurate.

DEFINITION: The *standard error of estimate* is a measure of dispersion around the line of regression. The symbol for the standard error of estimate is S_y.

FORMULA:

$$S_y = \sqrt{\frac{\Sigma (Y_c - Y)^2}{n}}$$

THE LOGIC OF THE FORMULA:

Given a line of regression for two variables, depicted in Figure 8.6.

To derive the standard error of estimate it is necessary to perform the following computations:

1. Calculate the deviation of each point, Y, from Y_c (computed Y).
2. Square each deviation, and sum the squares.
3. Divide the total by n, the number of items.
4. Compute the square root.

APPLICATION:

The standard error of estimate is used in the same manner as the standard deviation. The arithmetic mean, μ, plus or minus one standard

TABLE 8.1 PERCENTAGE OF POINTS INCLUDED WITHIN SPECIFIED STANDARD ERRORS OF ESTIMATE

Number of standard errors of estimate (S_y)	Percentage of points included
± 0.6745	50.00%
± 1.0000	68.27
± 1.6450	90.00
± 1.9600	95.00
± 2.0000	95.45
± 2.5760	99.00
± 3.0000	99.73

deviation, σ, includes 68.27 per cent of the items. Similarly, Y_c, or Y computed from the line of regression, plus or minus one standard error of estimate, S_y, includes 68.27 per cent of the points on the scatter diagram. This is based upon the assumption that the regression equation actually calculated from the sample observations is the true population regression equation, or that no errors are in the estimated values of the regression coefficients.

The confidence interval relationship is illustrated in Figure 8.7.

The z values and probabilities are the same as specified confidence intervals in relation to the normal curve. These may be summarized in Table 8.1.

Thus, use of the standard error of estimate enables the statistician to state with a specified degree of certainty how good his estimate is. If he predicts a value of the dependent variable, Y, from a given value of the independent variable, X, by using the line of regression, the statistician can also make a probability statement. Such a statement would give the confidence interval, or the lower and upper limits of his estimate, and the chances that the actual figure would fall within these limits.

8.7 THE COEFFICIENT OF CORRELATION

THE CONCEPT: The coefficient of correlation is a measure of the degree, or strength, of the relationship between the X and Y variables. This measurement of association represents explained variation, or the variation in the Y variable explained by variation in the X variable. To derive the value of explained variation, or degree of association, it is necessary to subtract the proportion of unexplained variance from 1. The square root

TABLE 8.2 SYMBOLS AND TERMS FOR VARIANCE AND VARIATION AS APPLIED TO LINEAR REGRESSION AND CORRELATION

Item	Variance	Variation
Total	σ_y^2	σ_y
Unexplained	S_y^2	S_y
Explained	$S_{y_c}^2$	S_{y_c}

of this result yields the coefficient of correlation, for which the symbol is r. Thus,

$$r^2 = 1 - \frac{\text{unexplained variance}}{\text{total variance}}$$

Consequently,

$$r = \sqrt{1 - \frac{\text{unexplained variance}}{\text{total variance}}}$$

As indicated in the previous chapter, *variance* is the square of the standard deviation. It is a number such that if total variance were 16 with explained variance = 12, and unexplained variance = 4, then the per cent of explained variance plus the per cent of unexplained variance = 100 per cent. In this illustration,

$$\text{explained variance} = \frac{12}{16} = \frac{3}{4} = 75\%$$

$$\text{unexplained variance} = \frac{4}{16} = \frac{1}{4} = 25\%$$

$$\text{total variance} = 75\% + 25\% = 100\%$$

Table 8.2 summarizes the symbols and terms for variance and variation as applied to linear regression and correlation. Variation is the square root of the variance.

For a graphic explanation of these terms, see Figure 8.8.

DEFINITION: *The coefficient of correlation* expresses the degree of association between the X and Y variables as a numerical value ranging from +1.00 to −1.00. The symbol for the coefficient of correlation is r.

With a range in values from +1.00 to −1.00, a zero value means "no correlation." The sign, plus (+) or minus (−), which must precede r, must be the same as the sign of the b value in the formula for the line of regression.

The r value in a particular problem must be either positive or negative. The sign, + or −, is assigned to r after it is computed.

The b value in the line-of-regression formula is called the *coefficient of regression* or the *coefficient of slope*. The b value is the *slope* of the regression line, that is, the change in Y per unit change in X. If b is a positive number, the line of regression slopes upward from left to right, and if b is negative, the line slopes downward from left to right.

It is interesting to note that the concepts of estimation and hypothesis testing apply also to the calculation of b in the linear equation of regression. The calculation of b is an example of point estimation. Hypothesis testing can be illustrated by the decision to accept or reject the value of calculated b as well as the value of calculated r. For example, the null hypothesis may be that b (or r) is equal to zero, and the alternative hypothesis that b (or r) is not equal to zero. It should be pointed out that b is distributed according to the t distribution.

Moreover, it is possible to compute the standard error of b, the *coefficient of regression*, as well as a, the Y intercept. In statistical work related to economics, the standard errrors of the regression coefficients are generally reported. They are used as any standard error measure: How much confidence may be placed upon an estimate of b, assuming that repeated samples may be drawn? The standard error of b is also useful if an alternative test of the null hypothesis of no correlation is desired.

The reliability of the measure, r, will depend upon the size of the sample. The larger the sample, the more reliable will be the value of r.

FORMULA:

$$r = \sqrt{1 - \frac{s_y^2}{\sigma_y^2}}$$

INTERPRETATION OF r: Table 8.3 provides verbal interpretations for the range of r values.

TABLE 8.3 DEGREE OF ASSOCIATION FOR THE RANGE OF r VALUES

Range of r	Degree of association
0 to ± 0.19	None or almost none
± .20 to ± 0.39	Low
± .40 to ± 0.59	Moderate
± .60 to ± 0.79	Marked
± .80 to ± 1.00	High

8.8 THE COEFFICIENT OF DETERMINATION

THE CONCEPT: The *coefficient of determination* measures the proportion of total variance which has been explained. It may be referred to as *explained variance*, and may be computed by squaring r, the coefficient of correlation.

DEFINITION: The *coefficient of determination* indicates the percentage of variation in the Y, or dependent, variable which has been explained by variation in the X, or independent, variable. The symbol is r^2.

FORMULA:

$$r^2 = \frac{\text{explained variance}}{\text{total variance}}$$

or

$$r^2 = \frac{S_{yc}^2}{\sigma_y^2}$$

or

$$r^2 = (r)(r)$$

EXAMPLE: If r, the coefficient of correlation for a particular problem, is $+ .91$ (high positive correlation), then $r^2 = 83$ per cent, as follows:

$$(.91)(.91) = .8281$$

and .8281 is transformed into a percentage by multiplying it by 100. The result means that 83 per cent of the variation in Y has been explained by variation in X.

8.9 THE COEFFICIENT OF NONDETERMINATION

THE CONCEPT: The *coefficient of nondetermination* measures the proportion of total variance that has not been explained.

DEFINITION: The *coefficient of nondetermination* indicates the percentage of variation in the Y, or dependent, variable which has *not* been explained by variation in the X, or independent, variable. The symbol is k^2.

FORMULA:

$$k^2 = \frac{\text{unexplained variance}}{\text{total variance}}$$

or

$$k^2 = \frac{S_y^2}{\sigma_y^2}$$

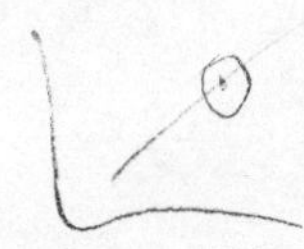

Given point Y on a scatter diagram (see Figure 8.8):

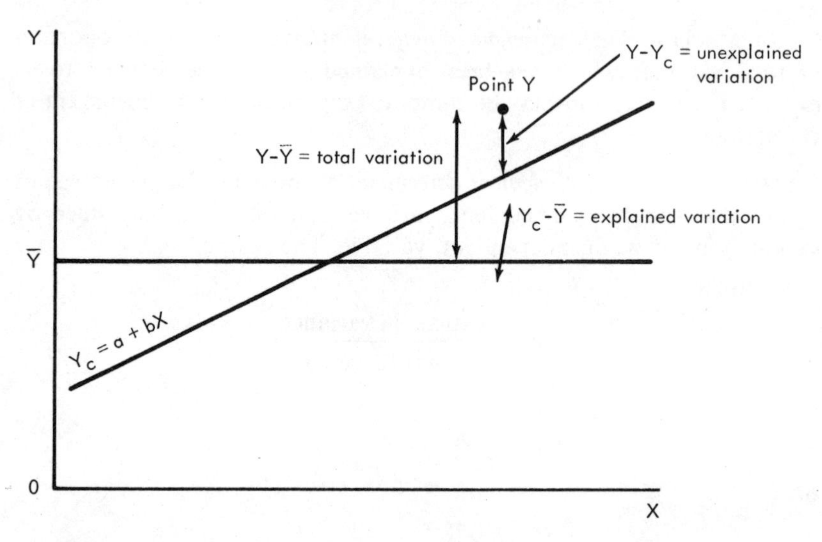

Fig. 8.8 Graphic Explanation of Explained, Unexplained, and Total Variation

Explained variation ($Y_c - \bar{Y}$): The amount of deviation that can be explained in terms of the X variable and predicted by the regression line equation, $Y_c = a + bX$.

Unexplained variation ($Y - Y_c$): A residual amount of deviation which cannot be predicted by knowing X and, therefore, is independent of X.

Total variation ($Y - \bar{Y}$): The amount of deviation by which an observed value differs from the mean of the Y's or $\bar{Y}$.

The *coefficient of determination* plus the *coefficient of nondetermination* equals 1, or

$$r^2 + k^2 = 1.0$$

However, $r + k$ is greater than 1.0 unless $r = 1.0$ or 0. If $r = 1$, the data are perfectly correlated, and $r + k$ would equal 1; $r + k$ would equal 1 also under conditions of no correlation when $r = 0$. The *coefficient of alienation* is called k. It measures the degree to which the X and Y variables are not related.

EXAMPLE: If $r^2 = 83$ per cent, then 100 per cent − 83 per cent = 17 per cent, or k^2.

The k^2 value of 17 per cent indicates that of the total variance (100 per cent), 17 per cent has not been explained by variation in X. This percentage represents the influence of other factors or forces which relate to variations in Y other than the influence of the X variable.

8.10 SUMMARY OF FORMULAS FOR LEAST SQUARES METHOD

Concept Formulas

1. Total variance: $\sigma_y^2 = \Sigma (Y - \bar{Y})^2/n$.
2. Explained variance: $S_{y_c}^2 = \Sigma (Y_c - \bar{Y})^2/n$.
3. Unexplained variance: $S_y^2 = \Sigma (Y_c - Y)^2/n$.

Total variance = explained variance + unexplained variance or

$$\sigma_y^2 = S_{y_c}^2 + S_y^2 .$$

Computational Forms

1. The line of regression is given by: $Y_c = a + bX$. The a and b values of the line of regression may be derived in either one of two ways:

a. Solve the two normal equations simultaneously. Appendix C indicates the procedure for this computation. The two normal equations are:

$$\text{(I)} \quad \Sigma Y = na + b\Sigma X$$
$$\text{(II)} \quad \Sigma XY = a\Sigma X + b\Sigma X^2$$

b. The a and b values may be computed directly from the following equations:

$$b = \frac{n(\Sigma XY) - (\Sigma X)(\Sigma Y)}{n(\Sigma X^2) - (\Sigma X)^2} \quad \text{or} \quad b = \frac{\Sigma XY - n\bar{X}\bar{Y}}{\Sigma X^2 - n\bar{X}^2}$$

After b has been computed, a may be derived by substituting b in one of the following equations:

$$a = \bar{Y} - b\bar{X} \quad \text{or} \quad a = \frac{\Sigma Y}{n} - b\frac{\Sigma X}{n}$$

2. Standard error of estimate (unexplained variation):

$$S_y = \sqrt{\frac{\Sigma (Y_c - Y)^2}{n}}$$

or

$$S_y = \sqrt{\frac{\Sigma Y^2 - a\Sigma Y - b\Sigma XY}{n}}$$

3. Coefficient of correlation

$$r = \sqrt{1 - \frac{S_y^2}{\sigma_y^2}}$$

where

$$\sigma_y^2 = \frac{\Sigma\, Y^2}{n} - \left(\frac{\Sigma\, Y}{n}\right)^2$$

or

$$r = \sqrt{\frac{a\,\Sigma\, Y + b\,\Sigma\, XY - \overline{Y}\,\Sigma\, Y}{\Sigma\, Y^2 - \overline{Y}\,\Sigma\, Y}}$$

or

$$r = \sqrt{\frac{a\,\Sigma\, Y + b\,\Sigma\, XY - n\left(\frac{\Sigma\, Y}{n}\right)^2}{\Sigma\, Y^2 - n\left(\frac{\Sigma\, Y}{n}\right)^2}}$$

4. Explained variance

$$S_{y_c}^2 = \frac{\Sigma\, Y_c^2}{n} - \left(\frac{\Sigma\, Y}{n}\right)^2$$

or

$$r^2 = (r)(r)$$

5. Explained variation

$$S_{y_c} = \sqrt{\frac{\Sigma\, Y_c^2}{n} - \left(\frac{\Sigma\, Y}{n}\right)^2}$$

6. Unexplained variance

$$S_y^2 = \frac{\Sigma\,(Y_c - Y)^2}{n}$$

or

$$k^2 = \frac{S_y^2}{\sigma_y^2}$$

Computational form for unexplained variation (standard error of estimate) is shown as formula 2.

7. Total variance

$$\sigma_y^2 = \frac{\Sigma\, Y^2}{n} - \left(\frac{\Sigma\, Y}{n}\right)^2$$

8. Standard deviation

$$\sigma_y = \sqrt{\frac{\sum Y^2}{n} - \left(\frac{\sum Y}{n}\right)^2}$$

Steps in Linear Regression and Correlation Analysis

1. Prepare a scatter diagram.
2. From the scatter diagram, ascertain the type of relationship between the variables, such as linear, and if it is linear, whether it is positive or negative.
3. Compute the basic measures describing the linear relationship.
4. Make the estimates or forecasts required.
5. Set up confidence intervals to describe the degree of precision.

8.11 LINEAR CORRELATION USING THE LEAST SQUARES METHOD

The data in Table 8.4 show newspaper advertising and sales for ten general merchandise stores in 1962:

TABLE 8.4 NEWSPAPER ADVERTISING AND SALES, GENERAL MERCHANDISE STORES, 1962
(millions of dollars)

Newspaper advertising (X)	Sales (Y)
2.1	10.5
3.6	20.1
5.9	33.4
0.8	4.7
0.4	2.7
8.6	52.5
0.2	1.9
1.0	4.3
4.2	27.1
7.7	55.4
34.5	212.6

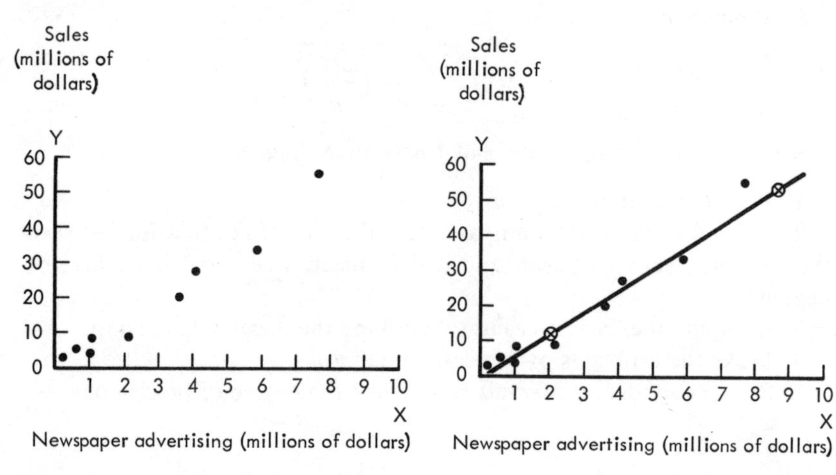

Fig. 8.9 Newspaper Advertising and Sales, Ten General Merchandise Stores, 1962

SOURCE: Table 8.4

Fig. 8.10 Newspaper Advertising and Sales, Ten General Merchandise Stores, 1962

1. Prepare a scatter diagram.
2. Ascertain the type of relationship between the variables by studying the scatter diagram, and state whether it is positive or negative.
3. Compute the linear equation of regression using the normal equations.
4. Plot the line of regression on the scatter diagram.
5. Compute the standard error of estimate of Y using the least squares concept formula.
6. Check answer to question 5 by using an alternate formula.
7. Compute the coefficient of correlation.
8. Check answer to question 7 by using an alternate formula.
9. Estimate sales for a general merchandise store which spends \$5 million for newspaper advertising.
10. Using a 95 per cent confidence interval, determine the limits of the estimate, and interpret the results.

Solution:

1. The scatter diagram appears in Figure 8.9.
2. From the scatter diagram it appears that the relationship is linear and positive. The points on the graph make a path from the lower left to the upper right.
3. Computation of the linear equation of regression using the normal equations (the necessary sums and products are shown in Table 8.5):

TABLE 8.5 NEWSPAPER ADVERTISING AND SALES, TEN GENERAL MERCHANDISE STORES, 1962: CALCULATION OF THE LINEAR REGRESSION EQUATION

(millions of dollars)

Newspaper advertising, X (1)	Sales, Y (2)	XY (3)	X^2 (4)
2.1	10.5	22.05	4.41
3.6	20.1	72.36	12.96
5.9	33.4	197.06	34.81
0.8	4.7	3.76	0.64
0.4	2.7	1.08	0.16
8.6	52.5	451.50	73.96
0.2	1.9	0.38	0.04
1.0	4.3	4.30	1.00
4.2	27.1	113.82	17.64
7.7	55.4	426.58	59.29
$\Sigma X = 34.5$	$\Sigma Y = 212.6$	$\Sigma XY = 1{,}292.89$	$\Sigma X^2 = 204.91$

$$n = 10$$

$$\bar{X} = \frac{\Sigma X}{n} = \frac{34.5}{10} = 3.45 \qquad \bar{Y} = \frac{\Sigma Y}{n} = \frac{212.6}{10} = 21.26$$

$$\text{(I)}\quad \Sigma Y = na + b\Sigma X$$

$$\text{(II)}\quad \Sigma XY = a\Sigma X + b\Sigma X^2$$

$$212.6 = 10a + 34.5b \leftarrow 3.45$$

$$1{,}292.89 = 34.5a + 204.91b$$

$$-733.47 = -34.5a - 119.025b$$

$$1{,}292.89 = 34.5a + 204.91b$$

$$559.42 = 85.885b$$

$$b = 6.51359$$

Substitute the b value in normal equation I:

$$212.6 = 10a + (34.5)(6.51359)$$

$$212.6 = 10a + 224.7189$$

$$-10a = 224.7189 - 212.6$$

$$-10a = +12.1189$$

$$a = -1.21189$$

Therefore, the linear regression line, $Y_c = a + bX$, is

$$Y_c = -1.21189 + 6.51359X$$

4. Plot the line of regression on the scatter diagram. Only two points are required, and therefore any two X values and their corresponding Y_c values may be used. For example, if $X = 2.1$ and $X = 8.6$, the equivalent Y_c values are derived as follows:

If $X = 2.1$:

$$\begin{aligned} Y_c &= -1.21189 + 6.51359X \\ &= -1.21189 + (6.51359)(2.1) \\ &= -1.21189 + 13.6785 \\ &= 12.4665 \end{aligned}$$

If $X = 8.6$:

$$\begin{aligned} Y_c &= -1.21189 + 6.51359X \\ &= -1.21189 + (6.51359)(8.6) \\ &= -1.21189 + 56.0169 \\ &= 54.8049 \end{aligned}$$

Thus, if $X = 2.1$, $Y_c = 12.4665$, and if $X = 8.6$, $Y_c = 54.8049$, the required two points (2.1, 12.4665) and (8.6, 54.8049) are plotted on the scatter diagram and connected by a straight line. This is illustrated in Figure 8.10. In this illustration, the line of regression was plotted by connecting the following two points: (2.1, 12.4665) and (8.6, 54.8049). These are indicated by ⊗.

5. Compute the standard error of estimate of Y using the least squares concept formula. The formula is $S_y = \sqrt{\Sigma (Y_c - Y)^2/n}$. To obtain $\Sigma (Y_c - Y)^2$, Table 8.5 is expanded as shown in Table 8.6. Substituting in

the preceding equation:

$$S_y = \sqrt{\frac{\Sigma (Y_c - Y)^2}{n}}$$

$$= \sqrt{\frac{77.409378}{10}}$$

$$= \sqrt{7.7409378}$$

$$= 2.7822, \quad \text{or} \quad \$2{,}782{,}000.$$

6. Check answer to question 5 by using an alternate formula. The following formula may be used:

$$S_y = \sqrt{\frac{\Sigma Y^2 - a\Sigma Y - b\Sigma XY}{n}}$$

$$= \sqrt{\frac{8{,}241.12 - (-1.21189)(212.6) - (6.51359)(1{,}292.89)}{10}}$$

$$= \sqrt{\frac{8{,}241.12 + 257.647814 - 8{,}421.3553751}{10}}$$

$$= \sqrt{\frac{77.412439}{10}}$$

$$= \sqrt{7.7412439}$$

$$= 2.7823, \quad \text{or} \quad \$2{,}782{,}000$$

TABLE 8.6 NEWSPAPER ADVERTISING AND SALES, TEN GENERAL MERCHANDISE STORES, 1962: DERIVATION OF $\Sigma (Y_c - Y)^2$

X	Y	XY	X^2	Y^2	Y_c	$Y_c - Y$	$(Y_c - Y)^2$
2.1	10.5	22.05	4.41	110.25	12.466649	1.966649	3.867708
3.6	20.1	72.36	12.96	404.01	22.237034	2.137034	4.566914
5.9	33.4	197.06	34.81	1,115.56	37.218291	3.818291	14.579346
0.8	4.7	3.76	0.64	22.09	3.998982	−0.701018	0.491426
0.4	2.7	1.08	0.16	7.29	1.393546	−1.306454	1.706822
8.6	52.5	451.50	73.96	2,756.25	54.804984	2.304984	5.312951
0.2	1.9	0.38	0.04	3.61	0.090828	−1.809172	3.273103
1.0	4.3	4.30	1.00	18.49	5.301700	1.001700	1.003403
4.2	27.1	113.82	17.64	734.41	26.145188	−0.954812	0.911666
7.7	55.4	426.58	59.29	3,069.16	48.942753	−6.457247	41.696039
34.5	212.6	1,292.89	204.91	8,241.12	212.599955	−0.000045	77.409378

NOTE: Y_c is computed from the line of regression formula, $Y_c = -1.21189 + 6.51359X$ by substituting each of the X values listed in the X column for X in this equation. For example, if $X = 2.1$, then $Y_c = 12.466649$, as indicated previously.

7. Compute the coefficient of correlation. The following formula may be used:

$$r = \sqrt{1 - \frac{S_y^2}{\sigma_y^2}}$$

where

$$\sigma_y^2 = \frac{\Sigma Y^2}{n} - \left(\frac{\Sigma Y}{n}\right)^2$$

σ_y^2 is obtained first as follows:

$$\begin{aligned}\sigma_y^2 &= \frac{8{,}241.12}{10} - \left(\frac{212.6}{10}\right)^2 \\ &= \frac{8{,}241.12}{10} - (21.26)^2 \\ &= 824.112 - 451.9876 \\ &= 372.1244\end{aligned}$$

Then,

$$\begin{aligned}r &= \sqrt{1 - \frac{7.7409378}{372.1244}} \\ &= \sqrt{1 - .0207} \\ &= \sqrt{.9703} \\ &= +.98\end{aligned}$$

8. Check answer to question 7 by using an alternate formula. The following formula may be used:

$$\begin{aligned}r &= \sqrt{\frac{a\Sigma Y + b\Sigma XY - \bar{Y}\Sigma Y}{\Sigma Y^2 - \bar{Y}\Sigma Y}} \\ &= \sqrt{\frac{(-1.21189)(212.6) + (6.51359)(1{,}292.89) - (21.26)(212.6)}{8{,}241.12 - (21.26)(212.6)}} \\ &= \sqrt{\frac{-257.647814 + 8{,}421.3553751 - 4{,}519.876}{8{,}241.12 - 4{,}519.876}} \\ &= \sqrt{\frac{+3{,}643.8315611}{3{,}721.244}} \\ &= \sqrt{.9792} \\ &= +.98\end{aligned}$$

9. Estimate sales for a general merchandise store which spends \$5

million for newspaper advertising. The line-of-regression formula is used:

$$\begin{aligned} Y_c &= -1.21189 + 6.51359X \\ &= -1.21189 + (6.51359)(5) \\ &= -1.21189 + 32.56795 \\ &= 31.35606, \quad \text{or} \quad \$31{,}356{,}000 \end{aligned}$$

Thus, a general merchandise store which spends $5 million for newspaper advertising may expect about $31,356,000 in sales.

10. Using a 95 per cent confidence interval, determine the limits of the estimate made in answer to question 9 and interpret the results.

$$\$31{,}356{,}000 \pm (1.96)(\$2{,}782{,}000)$$

$$\$31{,}356{,}000 \pm \$5{,}453{,}000$$

$$\begin{array}{rr} \$31{,}356{,}000 & \$31{,}356{,}000 \\ -\ \ 5{,}453{,}000 & +\ \ 5{,}453{,}000 \\ \hline \$25{,}903{,}000 & \$36{,}809{,}000 \end{array}$$

The limits are $25,903,000 to $36,809,000.

INTERPRETATION: The probability statement for these limits is that chances are 95 out of a 100 that sales will be between $25,903,000 and $36,809,000, when $5 million of newspaper advertising is spent.

The assumption made in this example is that the regression coefficients are the correct estimates of the true population regression coefficients.

8.12 THE PRODUCT MOMENT METHOD

THE CONCEPT: The product moment method provides a technique which reduces the amount of computation involved in obtaining measures of linear correlation. While the least squares method measures variation from the line of best fit, or the regression line, the product moment method measures variation from the point of means, or the intersection of the arithmetic means of the X's and Y's. The point of intersection between $\bar{X}$ and $\bar{Y}$ is shown in Figure 8.11.

APPLICATION: The product moment method provides a technique for computing measures of regression and correlation from the point of intersection of the arithmetic mean of the X's and the arithmetic mean of the Y's.

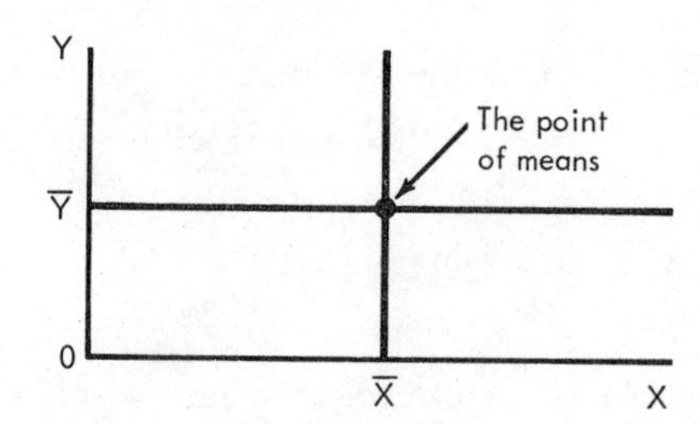

Fig. 8.11 The Product Moment Method: The Point of Means

FORMULAS: Computation of the basic measures in linear correlation is accomplished in the following order:

1. The coefficient of correlation, r
2. The standard error of estimate, S_y
3. The line of regression $Y_c = a + bX$.

Coefficient of Correlation, r

$$r = \frac{\Sigma\, xy}{\sqrt{(\Sigma\, x^2)(\Sigma\, y^2)}}$$

where

$$x = X - \bar{Y}$$

$$y = Y - \bar{Y}$$

It is important to note the distinction between small x and capital X, and small y and capital Y. Capital X represents the value of each item of the X series, and small x is the algebraic difference between each X value and the arithmetic mean of the X's.

Similarly, Y represents the value of each item of the Y series, and small y is the algebraic difference between each Y value and the arithmetic mean of the Y's.

Standard Error of Estimate, S_y

$$S_y = \sigma_y \sqrt{1 - r^2}$$

where

$$\sigma_y = \sqrt{\frac{\Sigma\, y^2}{n}}$$

The Line of Regression, $Y_c = a + bX$

$$Y_c = \bar{Y} + b\,(X - \bar{X})$$

where

$$b = \frac{\Sigma\, xy}{\Sigma\, x^2}$$

X in this formula is the same capital X that appears in the formula, $Y_c = a + bX$. To derive this regression line formula, numerical values are substituted only for $\bar{Y}$, b, and $\bar{X}$. These substitutions and solution yield the line of regression in the form, $Y_c = a + bX$.

8.13 LINEAR CORRELATION USING THE PRODUCT MOMENT METHOD

Using the same data as for the problem in Section 8.11 compute the following measures by the product moment method:

1. The coefficient of correlation
2. The standard error of estimate
3. The regression line

In Table 8.7, the first two columns, X and Y, are given. It is important to note the distinctions between capital X and capital Y and small x and small y.

TABLE 8.7 NEWSPAPER ADVERTISING AND SALES, TEN GENERAL MERCHANDISE STORES, 1962: PRODUCT MOMENT METHOD (millions of dollars)

Newspaper advertising X	Sales Y	x $(X - \bar{X})$	y $(Y - \bar{Y})$	xy	x^2	y^2
2.1	10.5	−1.35	−10.76	+14.5260	1.8225	115.7776
3.6	20.1	+0.15	−1.16	−0.1740	0.0225	1.3456
5.9	33.4	+2.45	+12.14	+29.7430	6.0025	147.3796
0.8	4.7	−2.65	−16.56	+43.8840	7.0225	274.2336
0.4	2.7	−3.05	−18.56	+56.6080	9.3025	344.4736
8.6	52.5	+5.15	+31.24	+160.8860	26.5225	975.9376
0.2	1.9	−3.25	−19.36	+62.9200	10.5625	374.8096
1.0	4.3	−2.45	−16.96	+41.5520	6.0025	287.6416
4.2	27.1	+0.75	+5.84	+4.3800	0.5625	34.1056
7.7	55.4	+4.25	+34.14	+145.0950	18.0625	1,165.5396
34.5	212.6	0	0	559.4200	85.8850	3,721.2440

$$\bar{X} = \frac{\Sigma X}{n} \qquad \bar{Y} = \frac{\Sigma Y}{n}$$

$$= \frac{34.5}{10} \qquad = \frac{212.6}{10}$$

$$= 3.45 \qquad = 21.26$$

1. Computation of the coefficient of correlation, r:

$$r = \frac{\Sigma\, xy}{\sqrt{(\Sigma\, x^2)(\Sigma\, y^2)}}$$

$$= \frac{559.42}{\sqrt{(85.885)\,(3{,}721.244)}}$$

$$= \frac{559.42}{\sqrt{319{,}599.0409}}$$

$$= \frac{559.42}{565.331}$$

$$= +.9895$$

2. Computation of the standard error of estimate, S_y:

$$S_y = \sigma_y \sqrt{1 - r^2}$$

where

$$\sigma_y = \sqrt{\frac{\Sigma\, y^2}{n}}$$

Obtain the value of σ_y by substitution:

$$\sigma_y = \sqrt{\frac{3{,}721.244}{10}}$$

$$= \sqrt{372.1244}$$

$$= 19.2905$$

Therefore,

$$S_y = 19.2905 \sqrt{1 - (.9895)^2}$$

$$= 19.2905 \sqrt{1 - .979110}$$

$$= 19.2905 \sqrt{.020890}$$

$$= (19.2905)\,(0.144553)$$

$$= 2.788, \quad \text{or} \quad \$2{,}788{,}000$$

The difference between the \$2,782,000 obtained by the least squares method and the \$2,788,000 obtained by the product moment method is due to rounding.

3. Computation of the line of regression, $Y_c = a + bX$:

$$Y_c = \bar{Y} + b(X - \bar{X})$$

where

$$b = \frac{\Sigma\, xy}{\Sigma\, x^2}$$

The value of b is obtained:

$$b = \frac{559.42}{85.885}$$

$$= 6.51359$$

Therefore,

$$\begin{aligned} Y_c &= 21.26 + 6.51359(X - 3.45) \\ &= 21.26 + 6.51359X - 22.4595 \\ &= -1.2 + 6.51359X \end{aligned}$$

EXERCISES

1. Define or explain the following:

Coefficient of correlation	Perfect correlation
Coefficient of determination	Positive correlation
Coefficient of nondetermination	Predicted variable
Coefficient of regression	Predicting variable
Coefficient of slope	Regression analysis
Correlation analysis	Regression line
Explained variance	Scatter diagram
Explained variation	Spurious correlation
Least squares method	Standard error of estimate
Linear correlation	Total variance
Linear regression	Total variation
Negative correlation	Unexplained variance
No correlation	Unexplained variation
Normal equations	

2. Given the data in Table 8.8. Use least squares method.

TABLE 8.8 GROSS SALES AND END-OF-YEAR INVENTORY OF TEN RETAIL STORES, 1964
(millions of dollars)

Gross sales (X)	End-of-year inventory (Y)
11	2
40	8
4	1
12	3
21	4
34	6
6	1
4	1
54	9
2	1
188	36

a. Prepare a scatter diagram.
b. Ascertain the type of relationship between the variables by studying the scatter diagram, and state whether it is positive or negative.
c. Compute the linear equation of regression, using the normal equations.
d. Plot the line of regression on the scatter diagram.
e. Compute the standard error of estimate of Y, using the least squares concept formula.
f. Check computation in question (e) by using an alternate formula.
g. Compute the coefficient of correlation.
h. Check computation in question (g) by using an alternate formula.
i. Estimate the end-of-year inventory for a retail store having annual sales of \$25 million.
j. Using a 95 per cent confidence interval, determine the limits of the estimate made in answer to question (i), and interpret the results.

3. Using the data of Table 8.8, above, compute the following measures by the product moment method.
   a. The coefficient of correlation
   b. The standard error of estimate
   c. The regression line

4. Given the data in Table 8.9.

TABLE 8.9 ADVERTISING COSTS AND ANNUAL SALES FOR SIX CIGARETTE BRANDS
(millions of dollars)

Advertising expenditures	Annual sales
2	10
4	20
6	33
1	5
1	3
9	52
23	123

a. Compute the regression equation.
b. Estimate annual sales where $7,500,000 is invested in advertising.
c. Establish a 95 per cent confidence interval for the estimate made in answer to question (b) above.
d. Compute the coefficient of determination, and explain its significance.

9

Time Series Analysis

In previous chapters, statistical methods were concerned primarily with observations which were analyzed in terms of frequency of occurrence. Based upon frequency distributions, measures of location and measures of variation were computed. And, in subsequent chapters problems in statistical inference and hypothesis testing were considered. The time element or the order of the data was unimportant.

In time series analysis, however, time or order is fundamental to the problem under consideration. Essentially, problems in time series are concerned with variation over time. For example, time is basic in such series as gross national product, or employment, or changes in prices, production, sales, or profits.

The economist and the businessman are intimately concerned with time series data. They read reports in the daily press, in journals, magazines, and a variety of government publications in order to keep up with statistical time series. However, the economist's and businessmans' primary concern with time series is the constantly urgent desire to estimate and evaluate future activity. Because of this need, probably more statisticians are engaged in forecasting than in any other branch of statistics.

This chapter first defines time series *and* time series analysis, *citing examples. It then considers the reasons for studying time series, and the steps required to analyze a time series.*

The four components of each observation in a time series are:

1. *Secular trend*
2. *Seasonal variation*

3. *Cyclical movement*
4. *Irregular or random forces*

Each of these components is analyzed in detail, as well as the procedure for eliminating the influence of each component in order to study that component by itself or the remaining components.

While the discussion of this chapter is concerned primarily with economic data, the methods apply equally well to time series of other activities in the social sciences as well as in the pure sciences.

9.1 INTRODUCTION TO TIME SERIES ANALYSIS

DEFINITIONS: A *time series* is a set of observations, of which each figure applies to a specific time period, such as a day, a week, a month, or a year.

Time series analysis is the statistical technique for studying fluctuations in time series data.

EXAMPLES OF TYPICAL TIME SERIES

1. The population of the United States at successive decennial censuses
2. Nonagricultural employment of the United States measured monthly over a period of years
3. The number of shares of stock sold on the New York Stock Exchange daily for a five-year period

Although time series data are most generally economic, time series are available for other branches of knowledge. For example, weather stations record hourly temperature, daily rainfall, and wind velocity. Accident statistics are maintained on a daily basis, while public health authorities maintain records on mortality rates resulting from various diseases, as well as birth statistics, and net population increase.

9.2 REASONS FOR STUDYING TIME SERIES

1. By studying the past, it becomes possible to understand and evaluate how a time series has evolved.

2. By relating the past to the present, it is possible to estimate future observations through the process of forecasting. Economists and businessmen are interested in both *long-term* and *short-term* forecasts.

DEFINITIONS: A *short-term forecast* is one which encompasses a time period ranging from a few days, weeks, or months up to about five years.

A *long-term forecast* is one which encompasses a time period ranging from five to twenty years.

Forecasts are widely used by management as the basis for planning. Forecasting serves well in the accounting department, in procurement, industrial relations, production, and in sales. A sound forecast of production will help the firm achieve optimum use of its labor supply, materials, and capital equipment, including both machinery and physical plant. Sales forecasting guides management in setting sales quotas, in equalizing sales potential among the firm's territories, and in determining the advertising budget.

9.3 STEPS IN ANALYZING A TIME SERIES

The steps required to analyze a time series generally follow the pattern indicated below.

1. Edit the data.
2. Make preliminary adjustments, if required. These include
   a. Calendar variation adjustment
   b. Population adjustment
   c. Price adjustment
3. Analyze the components of a time series.

Edit the Data

It is essential that data be checked to insure that each observation is comparable with data for different time periods. For example, if a firm is developing a series on production of computers, the statistician must be aware that the computer of 1960 is significantly different from the computer of 1950 or of 1940. Appropriate adjustments must be made in the definition to insure that the data are homogeneous.

Make Preliminary Adjustments, If Required

CALENDAR VARIATION: In working with a series of data, it is sometimes necessary to make specific adjustments in the data even before analysis of the time series may begin. For example, because February has fewer days than any other month, a series of monthly data would always show February at a lower volume of activity. It is possible to analyze the *rate* of activity by making an adjustment which would allocate an equal proportion of days to each month. This may be accomplished by using the figures in Table 9.1. Each monthly figure is multiplied by the adjustment

TABLE 9.1 MONTHLY CALENDAR VARIATION: ADJUSTMENT FACTORS

Month	Ordinary year		Leap year	
	Days	Adjustment factor	Days	Adjustment factor
January	31	0.981183	31	0.983871
February	28	1.086310	29	1.051724
March	31	0.981183	31	0.983871
April	30	1.013889	30	1.016667
May	31	0.981183	31	0.983871
June	30	1.013889	30	1.016667
July	31	0.981183	31	0.983871
August	31	0.981183	31	0.983871
September	30	1.013889	30	1.016667
October	31	0.981183	31	0.983871
November	30	1.013889	30	1.016667
December	31	0.981183	31	0.983871
Total	365		366	
Average	30.41667		30.5	

SOURCE: Kermit O. Hansen and George J. Brabb, *Managerial Statistics*, 2nd ed, Englewood Cliffs, N.J., Prentice-Hall, 1961, p. 178.

factor indicated. A similar type of adjustment may be made for working days, if it is required by the problem.

If data, such as sales, for example, are reported on an annual basis, either total annual sales or average monthly sales for each year, the data would not require adjustment for calendar variation.

POPULATION ADJUSTMENT: Analysis of a time series over a long period of time may yield an upward trend, principally because population has increased. Data may be adjusted for population change by dividing a given or actual figure by an estimate of population for the same time period. This adjustment expresses each observation on a per capita basis as follows:

$$\frac{\text{Actual figure}}{\text{Population figure for the same period}}$$

PRICE ADJUSTMENT: If a time series is expressed in dollars, the statistician may wish to eliminate the influence of price change, and thus express the series in constant dollars. The process of adjusting a series for price is known as *deflation*, and the price index used in this manner is referred to as the *deflator*.

DEFINITION: *Deflation* is a statistical technique used to transform a monetary series from *current* dollars to *real* dollars and thus correct for changes in the value of the monetary unit over time. The adjusted series is referred to as one *in constant dollars* or in dollars of constant purchasing power.

Data may be adjusted for price change by dividing a given or actual figure by a price index figure for the same period. The price index generally used is the Consumer Price Index, prepared by the U.S. Department of Labor, Bureau of Labor Statistics, or one of its components, or the Wholesale Price Index, or one of its components. The selection of one or the other price index is dependent upon the dollar series to be adjusted. Chapter 10 is concerned with the subject of index numbers.

FORMULA FOR DEFLATION:

$$\frac{\text{Actual figure}}{\text{Price index figure for the same period}}$$

EXAMPLE: Elimination of the influence of price change from a series on average weekly earnings is shown in Table 9.2.

TABLE 9.2 DERIVATION OF REAL AVERAGE WEEKLY EARNINGS FOR PRODUCTION WORKERS IN MANUFACTURING, SELECTED YEARS, 1939–1962

Year (1)	Average weekly earnings (nominal wages) (2)	U.S. Consumer Price Index (1957–1959 = 100.0) (3)	Real average weekly earnings (constant dollars of 1957–1959 purchasing power)[a] (4)
1939	$23.86	48.4	$49.30
1943	$43.14	60.3	$71.54
1948	$54.14	83.8	$64.61
1950	$59.33	83.8	$70.80
1955	$76.52	93.3	$82.02
1960	$89.72	103.1	$87.02
1961	$92.34	104.2	$88.62
1962	$96.56	105.4	$91.61

[a] Real earnings (col. 4) = nominal wages (col. 2) ÷ Consumer Price Index (col. 3).

Analyze the Components of a Time Series

The four components of each observation are:

Component	*Symbol*
1. Secular trend	T or Y_c (computed Y)
2. Seasonal variation	S
3. Cyclical movement	C
4. Irregular or random forces	I

The assumption generally made in time series analysis is that the given or original observation (symbol: O or Y) is equal to the product of the four components.

FORMULA: $O = T \times S \times C \times I.$

9.4 PROCEDURE FOR ELIMINATING THE INFLUENCE OF A PARTICULAR COMPONENT

In order to analyze one or another of the components, it is necessary to *decompose* the series. This is usually referred to as *eliminating the influence of* a particular component by first computing it, and then dividing the original observation by that component.

EXAMPLES

1. Elimination of the influence of trend: First compute the trend values, and then divide each original observation by the trend value for that period.

$$\frac{T \times S \times C \times I}{T} = S \times C \times I$$

2. Elimination of the influence of seasonal: First compute the seasonal index, and then divide each original observation by the seasonal value for that period.

$$\frac{T \times S \times C \times I}{S} = T \times C \times I$$

3. Elimination of the influence of trend and seasonal: After trend and seasonal values are computed, multiply the two values for each period, and divide each original observation by the product of trend and seasonal.

$$\frac{T \times S \times C \times I}{T \times S} = C \times I$$

4. Elimination of the influence of trend and cycle: After trend and cycle values are computed, multiply the two values for each period, and divide each original observation by the product of trend and cycle.

$$\frac{T \times S \times C \times I}{T \times C} = S \times I$$

The values for the irregular or random influence (I) cannot be computed separately, and therefore cannot be isolated. The random influence always remains as a residual combined with one of the other three components.

9.5 SECULAR TREND

THE CONCEPT: Secular change implies regularity and continuity. Secular influences on a time series are those which result in long-term growth or decline. Sudden upward or downward changes are inconsistent with the concept of secular change.

DEFINITION: *Secular trend* is the smooth, regular, long-term movement of a statistical series.

METHODS FOR MEASURING SECULAR TREND: Three basic methods are available for measuring secular trend. These are

1. The graphic method
2. The moving average method
3. The least squares method

The Graphic Method

The first step in the graphic method is to draw a graph of the data. Then, after observing the graph, a line is drawn through the data to indicate what appears to be the long-term movement. The line may be drawn in freehand, or with the aid of a ruler, or with drafting equipment.

The Moving Average Method

DEFINITION: The *moving average* is a method of eliminating passing fluctuations in a time series, thus developing values which define secular trend. In the moving average method, irregular upward and downward fluctuations are smoothed out by the process of averaging. The procedure results in a clear indication of trend.

EXAMPLE 1: Given a series of values covering a period of years (columns 1 and 2 of Table 9.3), compute a three-year moving average.

TABLE 9.3 ILLUSTRATION OF A THREE-YEAR MOVING AVERAGE

Year (1)	Y (2)	3-year moving total (3)	3-year moving average (4)
1954	7		
1955	9	31	10.33
1956	15	43	14.33
1957	19	54	18.00
1958	20	58	19.33
1959	19	61	20.33
1960	22	66	22.00
1961	25	73	24.33
1962	26		

The method involves computation of a series of averages. For a three-year moving average, the first three observations (7, 9, and 15) are added, and the total, 31, shown in column 3, is placed opposite the middle value, 9. Then, 31 divided by 3 equals 10.33, the three-year moving average, given in column 4.

The first number, 7, is then dropped, and the next number, 19, is picked up. The three items for the second moving total are 9, 15, and 19, which total 43, and average 14.33.

This process is continued until all the observations have been included.

The same method is used for a five-year, a seven-year, or a nine-year moving average. Larger numbers of years may be used, if required.

EXAMPLE 2: Given a series of values covering a period of years (columns 1 and 2 of Table 9.4), compute a two-year moving average.

For a two-year moving average, successive averages are computed for two-year periods. Thus, $7 + 9 = 16$, which is placed between 7 and 9 in column 3. Then, $16 \div 2 = 8.0$, indicated in column 4. This is the preliminary two-year moving average, because it is *uncentered*, that is, it is located between two values. To *center* or adjust these averages so that they coincide with the years, a two-year moving total of the preliminary two-year moving average is computed, 8.0 + 12.0 = 20.0, indicated in column 5. The final two-year moving average, column 6, is derived by

TABLE 9.4 ILLUSTRATION OF A TWO-YEAR MOVING AVERAGE

Year (1)	*Y* (2)	2-year moving total (3)	Preliminary 2-year moving average (4)	2-year moving total of preliminary 2-year moving average (5)	Final 2-year moving average (6)
1954	7				
		16	8.0		
1955	9			20.0	10.00
		24	12.0		
1956	15			29.0	14.50
		34	17.0		
1957	19			36.5	18.25
		39	19.5		
1958	20			39.0	19.50
		39	19.5		
1959	19			40.0	20.00
		41	20.5		
1960	22			44.0	22.00
		47	23.5		
1961	25			49.0	24.50
		51	25.5		
1962	26				

dividing the column 5 figure by 2. This end result is known as the *centered* two-year moving average.

The second total in column 3 (24) is the sum of 9 and 15; 34 in column 3 is the sum of 15 and 19. This process is continued until all the observations have been included.

Columns 3 and 4 would be similarly derived for a four-year moving average, a six-year, or an eight-year moving average. However, to *center* a four-year, six-year, or eight-year moving average involves the same procedure as indicated in columns 5 and 6. That is, an x-year moving total is computed for an *uncentered* four-year, six-year, or eight-year moving average, to obtain column 5, and then each total is divided by the appropriate number of years to derive column 6.[1]

The Least Squares Method

This method is used most frequently in the computation of secular trend. If the trend is linear, the straight line equation, $Y_c = a + bX$, is used. This least squares line for a particular series is derived from the two normal equations:

$$\text{(I)} \quad \Sigma Y = na + b\Sigma X$$
$$\text{(II)} \quad \Sigma XY = a\Sigma X + b\Sigma X^2$$

Appendix C explains the technique for solving two normal equations simultaneously.

EXAMPLE 1: ODD NUMBER OF ITEMS, ORIGIN AT FIRST YEAR. Determine a linear relationship between X and Y by the method of least squares given the following data, Y:

Year	*Y*
1954	7
1955	9
1956	15
1957	19
1958	20
1959	19
1960	22

Solution: To solve the two normal equations simultaneously, the following totals are required: ΣX, ΣY, ΣX^2, and ΣXY.

Because the origin is at the first year, 1954, the *zero* value is entered in the X column, and the successive years are numbered from 1 through 6 as indicated in column 2 of Table 9.5. The rest of the table is completed as shown.

[1] The influence of the business cycle causes erratic upward and downward movements of the original data. To eliminate these erratic fluctuations in an economic time series, the moving average method is used. The number of years included for the computation of a moving average, that is, a two-year, three-year, four-year, or other moving average period, should equal the average length of the business cycle over that period of years or a multiple of that length. The application of this principle will successfully smooth out cyclical fluctuations, and thus yield a better measure of trend.

TABLE 9.5 LEAST SQUARES TREND LINE: ORIGIN AT FIRST YEAR

Year (1)	X (2)	Y (3)	X^2 (4)	XY (5)
1954	0	7	0	0
1955	1	9	1	9
1956	2	15	4	30
1957	3	19	9	57
1958	4	20	16	80
1959	5	19	25	95
1960	6	22	36	132
Total	21	111	91	403

$$\text{(I)} \quad \Sigma Y = na + b\Sigma X$$
$$\text{(II)} \quad \Sigma XY = a\Sigma X + b\Sigma X^2$$

$$111 = 7a + 21b \qquad -3$$
$$403 = 21a + 91b$$

$$-333 = -21a - 63b$$
$$403 = 21a + 91b$$

$$70 = 28b$$
$$b = \frac{70}{28} = 2.5$$

$$111 = 7a + 21b$$
$$111 = 7a + (21)(2.5)$$
$$111 = 7a + 52.5$$
$$-7a = -111 + 52.5$$
$$+7a = +58.5$$
$$a = 8.357$$

Therefore:

$$Y_c = a + bX$$
$$Y_c = 8.357 + 2.5X$$

Origin: June 30/July 1, 1954
X units: one year

EXAMPLE 2: ODD NUMBER OF ITEMS: ORIGIN AT MIDDLE YEAR. Given the same data as in example 1 above, determine a linear relationship between X and Y by the method of least squares.

Solution: To simplify the procedure, the origin value in the X column, zero, is assigned to the middle year (Table 9.6). Minus values are assigned to the years prior to the origin year, and plus values to the years following the origin year.

TABLE 9.6 LEAST SQUARES TREND LINE: ORIGIN AT MIDDLE YEAR

Year (1)	X (2)	Y (3)	X^2 (4)	XY (5)	
1954	−3	7	9	−21	
1955	−2	9	4	−18	−54
1956	−1	15	1	−15	
1957	0	19	0	0	
1958	+1	20	1	+20	
1959	+2	19	4	+38	+124
1960	+3	22	9	+66	
Total	0	111	28		+70

$$\text{(I)} \quad \Sigma Y = na + b\Sigma X$$
$$\text{(II)} \quad \Sigma XY = a\,\Sigma X + b\Sigma X.$$

$$\text{(I)} \quad 111 = 7a$$
$$7a = 111$$
$$a = 15.857$$

$$\text{(II)} \quad 70 = 28b$$
$$28b = 70$$
$$b = \frac{70}{28} = 2.5$$

Therefore:

$$Y_c = a + bX$$

or

$$Y_c = 15.857 + 2.5X$$

Origin: June 30/July 1, 1957
X units: one year

EXAMPLE 3: EVEN NUMBER OF ITEMS: ORIGIN BETWEEN THE TWO MIDDLE YEARS. Determine a linear relationship between X and Y by the method of least squares, given the following data:

Year	Y
1955	7
1956	9
1957	15
1958	19
1959	20
1960	19

Solution: With an even number of years, no middle year is available for use as an origin. Therefore, the origin is assigned between the two

TABLE 9.7 LEAST SQUARES TREND LINE: ORIGIN BETWEEN TWO MIDDLE YEARS

Year (1)	X (2)	Y (3)	X^2 (4)	XY (5)	
1955	−5	7	25	−35	
1956	−3	9	9	−27	−77
1957	−1	15	1	−15	
1958	+1	19	1	+19	
1959	+3	20	9	+60	+174
1960	+5	19	25	+95	
Total	0	89	70		+97

$$\text{(I)} \quad \Sigma Y = na + b\Sigma X$$
$$\text{(II)} \quad \Sigma XY = a\,\Sigma X + b\Sigma X^2$$

$$\text{(I)} \quad 89 = 6a$$
$$6a = 89$$
$$a = \frac{89}{6} = 14.833$$

$$\text{(II)} \quad 97 = 70b$$
$$70b = 97$$
$$b = \frac{97}{70} = 1.386$$

Therefore:

$$Y_c = a + bX$$

or

$$Y_c = 14.833 + 1.386X$$

Origin: December 31, 1957/January 1, 1958
X units: 1/2 year, or 6 months

middle years, December 31, 1957/January 1, 1958. X units on both sides of the origin are measured in 6-month periods in order to compute to the June 30/July 1 date for each. Therefore, the first year on both sides of the origin is one 6-month period away from the origin.

The second year on either side of the origin is 18 months away from the origin, or three 6-month periods. Thus, −3 and +3 values are assigned in the X column in Table 9.7.

The third year on either side of the origin is five 6-month periods away from the origin, and −5 and +5 values are assigned in the X column of Table 9.7.

9.6 SPECIAL PROCEDURES RELATING TO THE TREND EQUATION

Two special procedures relating to the trend equation are important. These are *transformation in the trend line* and *shifting the origin.*

Transformation in the Trend Line

THE CONCEPT: In making a complete analysis of a time series, it is generally necessary to work with monthly data. The volume of computation required to compute the trend equation from monthly data may be reduced if annual data also are available. The time series analyst first computes the annual trend equation and then transforms the trend line from annual data, for which X units equal one year, to monthly data, for which X units equal one month. This *transformation in the trend line* is the first special procedure relating to the trend equation.

There are two *types of annual data:* annual totals and monthly averages. The sum of the twelve months of data for a particular year is the annual total. The average of the twelve months is the monthly average. For example, the figures in Table 9.8 represent production of a particular

TABLE 9.8 MONTHLY AND ANNUAL PRODUCTION OF A PRODUCT

Month	Production
January	12
February	14
March	15
April	11
May	13
June	16
July	17
August	19
September	14
October	13
November	12
December	11
Total	168

$$\text{Arithmetic mean} = \frac{168}{12} = 14$$

product for one year. The annual total is 168, and the monthly average is 14. A series of annual data, therefore, may consist of annual totals or monthly averages. The particular series usually specifies the type of annual data given.

To simplify the process of transforming annual data to monthly data, the *transformation table* Table 9.9, may be used.

TABLE 9.9 PROCEDURES FOR TRANSFORMING ANNUAL STRAIGHT-LINE TREND EQUATION TO A MONTHLY TREND EQUATION

Number of years	Type of annual data			
	Monthly averages		Annual totals	
	a	*b*	*a*	*b*
Odd	No change	Divide by 12	Divide by 12	Divide by 144
Even	No change	Divide by 6	Divide by 12	Divide by 72

SOURCE: Frederick E. Croxton and Dudley J. Cowden, *Applied General Statistics*, 2nd ed, Englewood Cliffs, N.J., Prentice-Hall, 1955, p. 278.

Using the table requires prior determination of the type of annual data in the problem (monthly averages or annual totals) and the number of years in the series (odd number of years or even number). The *a* and *b* columns refer to the *a* and *b* constants in the linear trend equation ($Y_c = a + bX$), and the stub entry under each indicates the adjustment which must be made for each constant in the trend equation.

EXAMPLE: Transform the following trend equation (based upon annual totals of data for an odd number of years) to monthly data.

$$Y_c = 317.3 + 16.7X$$

Origin: June 30/July 1, 1950

X units: one year

Solution: Divide the *a* constant by 12. Divide the *b* constant by 144. Therefore:

$$Y_c = \frac{317.3}{12} + \frac{16.7X}{144}$$

$$Y_c = 26.44 + 0.116X$$

Origin: June 30/July 1, 1950

X units: one month

Shifting the Origin

The second special procedure is known as *shifting the origin*, which is the process of changing the starting point for computation of trend values. If it is desired to shift an origin either forward or backward in time, the following procedure is used.

STEPS

1. Count the number of months from the given origin to the new origin.
2. Multiply the figure obtained in step 1 by the b constant of the month-to-month equation.
3. Either of the following:

When shifting backward: Subtract the product obtained in step 2 from the a constant to obtain the adjusted a constant.

When shifting forward: Add the product obtained in step 2 to the a constant to obtain the adjusted a constant.

EXAMPLE: Given:

$$Y_c = 26.44 + 0.116X$$

Origin: June 30/July 1, 1950
X units: one month

Shift the origin backward to January 15, 1945.

Solution: Following the steps indicated above,

1. The number of months from June 30, 1950 to January 15, 1945 is 65.5.
2. (65.5) (0.116) = 7.598.
3. 26.44 − 7.598 = 18.84.

Therefore:

$$Y_c = 18.84 + 0.116X$$

Origin: January 15, 1945
X units: one month

9.7 SELECTING A CURVE TO REPRESENT TREND

In order to select the curve type, the data should be plotted in four different combinations. The first two relate directly to economic time series and will generally meet the requirements of the problem.

1. Natural X, natural Y. Arithmetic graph paper is used.
2. Natural X, log Y. Semilogarithmic paper is used, with X's plotted on the natural scale and Y's on the logarithmic scale.
3. Natural Y, log X. Semilogarithmic paper is used, with the X scale logarithmic.
4. Log Y, log X. Log-log paper is used. Both scales are logarithmic.

Personal judgment plays a major role in selecting the type of curve which best represents trend in a given problem. Experience is very important in making this judgment. If a straight-line trend is shown, an equation which plots as a straight line would be used. If a linear equation is inappropriate, the plotted data will suggest some other simple type, such as the parabola, hyperbola, or the exponential type.

The process of fitting a trend is a technique for approximating the average which underlies the long-run movement of a series. Mathematically, it is possible to secure any degree of smoothness in a time series. Generally, it is preferable to obtain the simplest curve possible. While linear trend has been emphasized, it is evident that more than one trend can be fitted to a given set of data, and an infinite number of shapes is possible.

9.8 A PROBLEM IN TIME SERIES ANALYSIS: SECULAR TREND

The figures in Table 9.10 indicate production of Portland cement in the United States for the years 1952–1962.

TABLE 9.10 PORTLAND CEMENT PRODUCTION IN THE UNITED STATES, 1952–1962
(millions of barrels)

Year	Portland cement production
1952	248.9
1953	263.8
1954	271.3
1955	296.8
1956	316.5
1957	297.8
1958	311.2
1959	338.5
1960	318.7
1961	323.4
1962	336.3

SOURCE: U.S. Department of Commerce, *Survey of Current Business*.

1. Using 1952 as the origin year, solve the normal equations for a and b, and set up the linear equation of trend. Cite the origin and X units.
2. Plot the original data and the trend equation on graph paper.
3. Using 1957 as the origin year, solve the normal equations, and set up the linear equation of trend. Cite the origin and X units.
4. Is the trend equation computed in question 3 the same as that computed in question 1? Explain.
5. Using the June 30/July 1, 1957, date as origin, transform the linear equation of trend from X units equal to one year to X units equal to one month.
6. Using the linear equation of trend, origin June 30/July 1, 1957, and X units equal to one month, shift the origin to January 15, 1952.
7. Using the annual trend equation, what is the trend value for the year 1965?
8. Using the month-to-month trend equation, what is the trend value for March 1963?

QUESTION 1. Using 1952 as the origin year, solve the normal equations for a and b, and set up the linear equation of trend. Cite the origin and X units.

Solution. Set up Table 9.11.

TABLE 9.11 LEAST SQUARES TREND LINE FOR PORTLAND CEMENT PRODUCTION, 1952–1962: ORIGIN AT FIRST YEAR

(production in millions of barrels)

Year	X	Portland cement production, Y (given)	XY	X^2
1952	0	248.9	0	0
1953	1	263.8	263.8	1
1954	2	271.3	542.6	4
1955	3	296.8	890.4	9
1956	4	316.5	1266.0	16
1957	5	297.8	1489.0	25
1958	6	311.2	1867.2	36
1959	7	338.5	2369.5	49
1960	8	318.7	2549.6	64
1961	9	323.4	2910.6	81
1962	10	336.3	3363.0	100
	55	3323.2	17511.7	385

SOURCE: Table 9.10.

Solve the two normal equations:

$$\begin{array}{lrl} \text{(I)} & \Sigma Y = & na \quad + b\Sigma X \\ \text{(II)} & \Sigma XY = & a\,\Sigma X + b\Sigma X^2 \\ \hline \text{(I)} & 3323.2 = & 11a + \ 55b \qquad -5 \\ \text{(II)} & 17511.7 = & 55a + 385b \\ \hline \text{(I)} & -16616.0 = & -55a - 275b \\ \text{(II)} & 17511.7 = & 55a + 385b \\ \hline & 895.7 = & 110b \\ & b = & 8.1427 \end{array}$$

Substitute b value in formula I:

$$\begin{aligned} 3323.2 &= 11a + (55)(8.1427) \\ 3323.2 &= 11a + 447.8485 \\ 11a &= 3323.2 - 447.8485 \\ 11a &= 2875.3515 \\ a &= 261.3956 \end{aligned}$$

Thus, the linear equation of trend is

$$Y_c = 261.3956 + 8.1427X$$

Origin: June 30/July 1, 1952
X units: one year

QUESTION 2. Plot the original data and the trend line on a sheet of graph paper.

Solution: In order to plot the trend line on graph paper, it is necessary to calculate the trend value (Y_c) for at least two years. Usually, the first and last years are selected. In this problem Y_c is computed for 1952 and 1962. Y_c for 1952 is derived as follows:

$$\begin{aligned} Y_c &= 261.3956 + 8.1427X \\ &= 261.3956 + (8.1427)(0) \\ &= 261.3956 \end{aligned}$$

Y_c for 1962 is derived as follows:

$$\begin{aligned} Y_c &= 261.3956 + 8.1427X \\ &= 261.3956 + (8.1427)(10) \\ &= 261.3956 + 81.4270 \\ &= 342.8226 \end{aligned}$$

The original data and the two required points for the trend line are plotted in Figure 9.1.

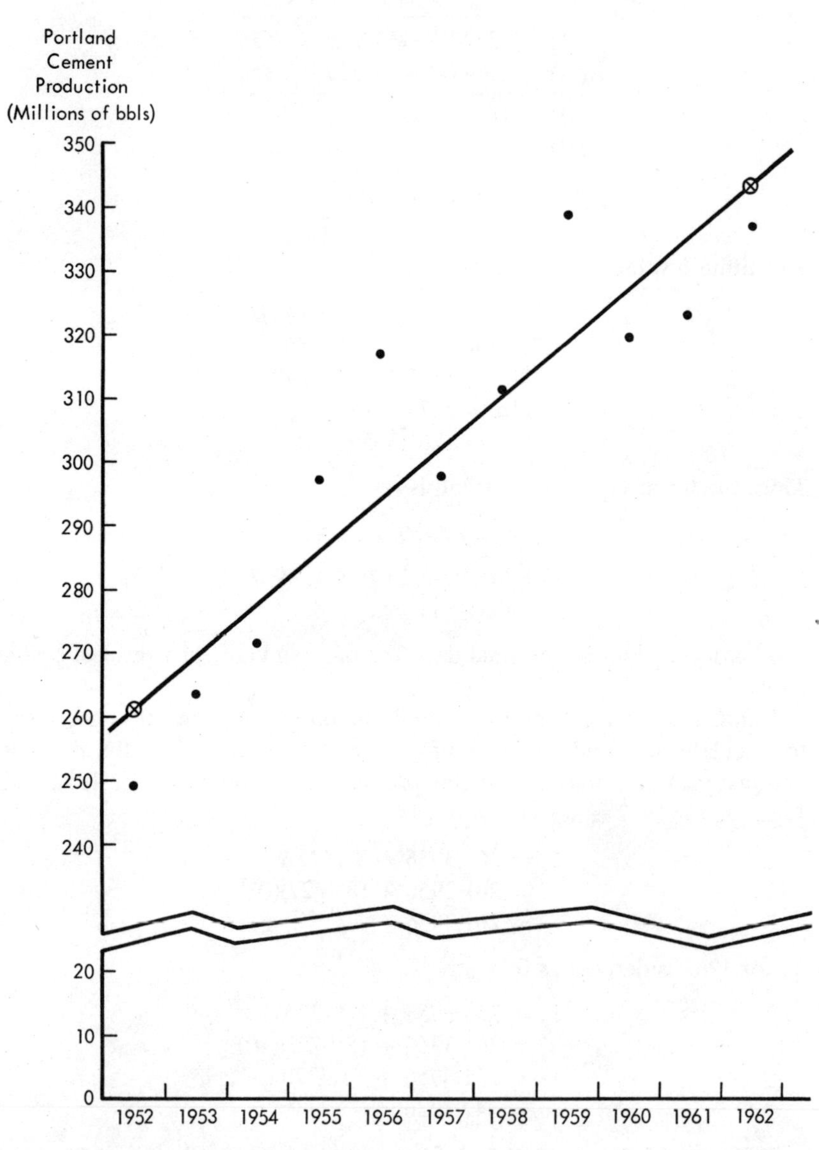

Fig. 9.1 Portland Cement Production in the United States, 1952–1962: Original Data and Trend Line

QUESTION 3. Using 1957 as the origin year, solve the normal equations and set up the linear equation of trend. Cite the origin and X units.
Solution: Set up Table 9.12.

TABLE 9.12 LEAST SQUARES TREND LINE FOR PORTLAND CEMENT PRODUCTION (Y), ORIGIN AT MIDDLE YEAR
(production in millions of barrels)

Year	X	Y	XY	X^2
1952	−5	248.9	−1244.5	25
1953	−4	263.8	−1055.2	16
1954	−3	271.3	−813.9	9
1955	−2	296.8	−593.6	4
1956	−1	316.5	−316.5	1
1957	0	297.8	0	0
1958	1	311.2	311.2	1
1959	2	338.5	677.0	4
1960	3	318.7	956.1	9
1961	4	323.4	1293.6	16
1962	5	336.3	1681.5	25
	0	3323.2	895.7	110

SOURCE: Table 9.10.

Solve the two normal equations:

$$\text{(I)} \quad \Sigma Y = na \quad + b\Sigma X$$
$$\text{(II)} \quad \Sigma XY = a\Sigma X + b\Sigma X^2$$

$$\text{(I)} \; 3323.2 = 11a$$
$$a = 302.1091$$

$$\text{(II)} \quad 895.7 = 110b$$
$$b = 8.1427$$

Thus, the linear equation of trend is:

$$Y_c = 302.1091 + 8.1427X$$

Origin: June 30/July 1, 1957
X units: one year

QUESTION 4. Is the trend equation computed in answer to question 3 above the same as that computed in question 1? Explain.

Solution: The two equations are the same. The a constant of Y intercept changes as the origin changes. However, the slope of the line is the same, and in spite of different a constants, if both equations were drawn on the same graph, the trend lines would be identical.

QUESTION 5. Using the June 30/July 1, 1957 date as origin, transform the linear equation of trend from X units equal to one year to X units equal to one month.

Solution: The data are annual totals and apply to an odd number of years, 11 years. Therefore, to transform annual data to monthly data, the a constant is divided by 12, and the b constant by 144 (see Table 9.9 page 193). Thus,

$$Y_c = \frac{302.1091}{12} + \frac{8.1427X}{144}$$
$$= 25.1758 + 0.0565X$$

Origin: June 30/July 1, 1957
X units: one month

QUESTION 6. Using the linear equation of trend, origin June 30/July 1, 1957, and X units equal to one month, shift the origin to January 15, 1952.

Solution: In order to shift the origin backward, it is necessary to multiply the b constant by the number of months involved, and to subtract the product from the a constant. The steps are as follows:

1. The number of months from June 30/July 1, 1957, back to January 15, 1952, is 65.5.
2. $(b)(65.5) = (0.0565)(65.5) = 3.7008$
3. $a - 3.7008 = 25.1758 - 3.7008 = 21.4750$

Thus, the linear equation of trend is

$$Y_c = 21.4750 + 0.0565X$$

Origin: January 15, 1952
X units: one month

QUESTION 7. Using the annual trend equation, what is the trend value for the year 1965?

Solution: Using the trend equation with origin June 30/July 1, 1957, and X units equal to one year, the first step is the calculation of the number of years to 1965. The X value is computed as 8. Therefore, 8 is substituted for X in the equation:

$$Y_c = 302.1091 + 8.1427X$$
$$= 302.1091 + (8.1427)(8)$$
$$= 302.1091 + 65.1416$$
$$= 367.2507$$

Therefore, the trend value for the year 1965 is 367.2507 millions of barrels.

QUESTION 8. Using the month-to-month trend equation, what is the trend value for March 1963?

Solution: Using the trend equation with origin June 30/July 1, 1957, and X units equal to one month, the first step is the calculation of the number of months to March 1963. The X value is computed as 69. Therefore, 69 is substituted for X in the equation:

$$\begin{aligned} Y_c &= 25.1758 + 0.0565X \\ &= 25.1758 + (0.0565)(69) \\ &= 25.1758 + 3.8985 \\ &= 29.0743 \end{aligned}$$

Therefore, the trend value for March 1963 is 29.0743 millions of barrels.

9.9 SEASONAL VARIATION

THE CONCEPT: In analyzing seasonal variation the end result is an index of seasonal fluctuations. The seasonal index indicates for the time series the typical performance for each month or quarter. It is, in effect, therefore, an average of activity for all the Januarys in the series, all the Februarys, all the Marchs, and each of the subsequent months; or for quarterly indexes, it is an average of all the first quarters, second quarters, and so forth.

DEFINITION. *Seasonal index* is the term describing month-to-month or quarter-to-quarter movements within a year that recur year after year. The seasonal index for month-to-month data consists of twelve number values, one for each month, whose twelve-month total is 1,200, and whose average is 100.0. For quarter-to-quarter movements, the seasonal index consists of four number values, one for each quarter, whose four-quarter total is 400, and whose average is 100.0.

9.10 USES OF THE SEASONAL INDEX IN TIME SERIES ANALYSIS

1. By computing the index of seasonal fluctuations for a time series, statisticians are able to eliminate the seasonal influence in the original data, thus isolating the value of the product of trend, cyclical, and irregular forces.

2. Having the seasonal index for a particular month enables the statistician to evaluate a given month's actual performance against past and present performance for the same month. The quarterly seasonal index makes possible a similar analysis for quarterly data.

3. Knowledge of the seasonal index is an essential tool in forecasting.

9.11 RATIO-TO-MOVING-AVERAGE METHOD FOR MEASURING SEASONAL VARIATION

A large number of methods are available for measuring seasonal variation. One of the most widely used techniques is the ratio-to-moving-average method.

Computing a Seasonal Index by the Ratio-to-moving-average Method

1. From the original data, compute the 12-month moving totals for the entire series.
2. Compute a 2-month moving total of the 12-month moving total.
3. Compute the recentered 12-month moving average by dividing the results obtained in step 2 by 24.
4. Compute the actual, or original, value for each month as a relative of the moving average by dividing each original value by the recentered 12-month moving average.
5. Prepare a multiple frequency distribution table, using the relatives computed in step 4.
6. Compute the sum of the middle set of relatives from the multiple frequency distribution table.
7. Compute the unadjusted arithmetic mean.
8. Adjust the monthly arithmetic means to equal 1,200. These adjusted arithmetic means constitute the index of seasonal variation.

9.12 A PROBLEM IN TIME SERIES ANALYSIS: THE SEASONAL INDEX

The Problem

1. Given the monthly data on Portland cement production in the United States, 1952–1962 (column 2 of Table 9.13), compute the seasonal index.
2. Plot the seasonal index on graph paper.

TABLE 9.13 PORTLAND CEMENT PRODUCTION IN THE UNITED STATES, MONTHLY, JANUARY 1952–DECEMBER 1962: CALCULATION OF PER CENT OF TWELVE-MONTH MOVING AVERAGE

Year and month (1)	Portland cement production (millions of barrels) (2)	12-month moving total (3)	2-month moving total of Col. 3 (4)	Recentered 12-month moving average (Col. 4 ÷ 24) (5)	Actual production as relative of moving average (Col. 2 ÷ Col. 5) (6)
1952 Jan	17.0				
Feb	16.5				
Mar	18.1				
Apr	19.8				
May	21.8				
Jun	20.7				
		248.9			
Jul	21.3		499.7	20.8	102.4
		250.8			
Aug	23.6		502.4	20.9	112.9
		251.6			
Sep	23.0		505.3	21.1	109.0
		253.7			
Oct	24.2		509.4	21.2	114.2
		255.7			
Nov	22.0		512.9	21.4	102.8
		257.2			
Dec	20.9		516.4	21.5	97.2
		259.2			
1953 Jan	18.9		521.2	21.7	87.1
		262.0			
Feb	17.3		524.7	21.9	79.0
		262.7			
Mar	20.2		526.2	21.9	92.2
		263.5			
Apr	21.8		527.5	22.0	99.1
		264.0			
May	23.3		528.5	22.0	105.9
		264.5			
Jun	22.7		528.3	22.0	103.2
		263.8			
Jul	24.1		526.5	21.9	110.0
		262.7			
Aug	24.3		525.0	21.9	111.0
		262.3			
Sep	23.8		524.5	21.9	108.7
		262.2			
Oct	24.7		524.3	21.8	113.3
		262.1			
Nov	22.5		524.2	21.8	103.2
		262.1			
Dec	20.2		524.3	21.8	92.7
		262.2			

(Table 9.13 continued)

TABLE 9.13 (CONTINUED)

Year and month (1)	Portland cement production (millions of barrels) (2)	12-month moving total (3)	2-month moving total of Col. 3 (4)	Recentered 12-month moving average (Col. 4 ÷ 24) (5)	Actual production as relative of moving average (Col. 2 ÷ Col. 5) (6)
1954 Jan	17.8	262.6	525.8	21.9	81.3
Feb	16.9	265.0	528.6	22.0	76.8
Mar	20.1	266.7	531.7	22.2	90.5
Apr	21.7	267.9	534.6	22.3	97.3
May	23.3	269.2	537.1	22.4	104.0
Jun	22.8	271.3	540.5	22.5	101.3
Jul	25.5	273.7	545.0	22.7	112.3
Aug	25.7	274.4	548.1	22.8	112.7
Sep	25.5	276.6	551.0	23.0	110.9
Oct	25.9	279.7	556.3	23.2	111.6
Nov	23.8	283.4	563.1	23.5	101.3
Dec	22.3	287.4	570.8	23.8	93.7
1955 Jan	20.2	282.9	576.6	24.0	84.2
Feb	17.6	291.4	580.6	24.2	72.7
Mar	22.3	292.9	584.3	24.3	91.8
Apr	24.8	294.9	587.8	24.5	101.2
May	27.0	296.0	590.9	24.6	109.8
Jun	26.8	296.8	592.8	24.7	108.5
Jul	27.3	298.0	594.8	24.8	110.1
Aug	27.9	300.0	598.0	24.9	112.1
Sep	27.0	301.1	601.1	25.0	108.0
Oct	27.9	302.4	603.5	25.1	111.2
Nov	24.9	305.0	607.4	25.3	98.4
Dec	23.1	307.0	612.0	25.5	90.6

(Table 9.13 continued)

TABLE 9.13 (CONTINUED)

Year and month (1)	Portland cement production (millions of barrels) (2)	12-month moving total (3)	2-month moving total of Col. 3 (4)	Recentered 12-month moving average (Col. 4 ÷ 24) (5)	Actual production as relative of moving average (Col. 2 ÷ Col. 5) (6)
1956 Jan	21.4		616.2	25.7	83.3
		302.9			
Feb	19.6		620.6	25.9	75.7
		311.4			
Mar	23.4		624.4	26.0	90.0
		313.0			
Apr	26.1		627.2	26.1	100.0
		314.2			
May	29.6		629.4	26.2	113.0
		315.2			
Jun	28.8		631.7	26.3	109.5
		316.5			
Jul	29.5		630.9	26.3	112.2
		314.4			
Aug	30.1		627.0	26.1	115.3
		312.6			
Sep	28.6		624.4	26.0	110.0
		311.8			
Oct	29.1		621.5	25.9	112.4
		309.7			
Nov	25.9		617.3	25.7	100.8
		307.6			
Dec	24.4		612.9	25.5	95.7
		305.3			
1957 Jan	19.3		601.4	25.1	76.9
		296.1			
Feb	17.8		593.5	24.7	72.1
		297.4			
Mar	22.6		597.1	24.9	90.8
		299.7			
Apr	24.0		600.4	25.0	96.0
		300.7			
May	27.5		600.5	25.0	110.0
		299.8			
Jun	26.5		597.6	24.9	106.4
		297.8			
Jul	20.3		594.5	24.8	81.9
		296.7			
Aug	31.4		589.7	24.6	127.6
		293.0			
Sep	30.9		581.4	24.2	127.7
		288.4			
Oct	30.1		576.8	24.0	125.4
		288.4			
Nov	25.0		578.6	24.1	103.7
		290.2			
Dec	22.4		584.0	24.3	92.2
		293.8			

(Table 9.13 continued)

TABLE 9.13 (CONTINUED)

Year and month (1)	Portland cement production (millions of barrels) (2)	12-month moving total (3)	2-month moving total of Col. 3 (4)	Recentered 12-month moving average (Col. 4 ÷ 24) (5)	Actual production as relative of moving average (Col. 2 ÷ Col. 5) (6)
1958 Jan	18.2		597.1	24.9	73.1
		303.3			
Feb	14.1		606.9	25.3	55.7
		303.6			
Mar	18.0		607.9	25.3	71.1
		304.3			
Apr	24.0		611.3	25.5	94.1
		307.0			
May	29.3		617.0	25.7	114.0
		310.3			
Jun	30.1		621.2	25.9	116.2
		311.2			
Jul	29.8		622.8	26.0	114.6
		311.6			
Aug	31.7		625.8	26.1	121.5
		314.2			
Sep	31.6		634.7	26.4	119.7
		320.5			
Oct	32.8		646.1	26.9	121.9
		325.6			
Nov	28.0		655.3	27.3	102.6
		329.7			
Dec	23.6		662.8	27.6	85.5
		333.1			
1959 Jan	18.6		670.6	27.9	66.7
		337.5			
Feb	16.7		678.1	28.3	59.0
		340.6			
Mar	24.3		682.2	28.4	85.6
		341.6			
Apr	29.1		681.5	28.4	102.5
		339.9			
May	33.4		677.9	28.2	118.4
		338.0			
Jun	33.5		676.5	28.2	118.8
		338.5			
Jul	34.2		677.1	28.2	121.3
		338.6			
Aug	34.8		676.6	28.2	123.4
		338.0			
Sep	32.6		670.1	27.9	116.8
		332.1			
Oct	31.1		662.1	27.6	112.7
		330.0			
Nov	26.1		658.4	27.4	95.3
		328.4			
Dec	24.1		655.2	27.3	88.3
		326.8			

(Table 9.13 continued)

TABLE 9.13 (CONTINUED)

Year and month (1)	Portland cement production (millions of barrels) (2)	12-month moving total (3)	2-month moving total of Col. 3 (4)	Recentered 12-month moving average (Col. 4 ÷ 24) (5)	Actual production as relative of moving average (Col. 2 ÷ Col. 5) (6)
1960 Jan	18.7		651.4	27.1	69.0
		324.6			
Feb	16.1		647.7	27.0	59.6
		323.1			
Mar	18.4		644.7	26.9	68.4
		321.6			
Apr	27.0		643.5	26.8	100.7
		321.9			
May	31.8		644.2	26.8	118.7
		322.3			
Jun	31.9		641.0	26.7	119.5
		318.7			
Jul	32.0		635.4	26.5	120.8
		316.7			
Aug	33.3		632.3	26.3	126.6
		315.6			
Sep	31.1		634.7	26.4	117.8
		319.1			
Oct	31.4		637.7	26.6	118.0
		318.6			
Nov	26.5		636.5	26.5	100.0
		317.9			
Dec	20.5		635.5	26.5	77.4
		317.6			
1961 Jan	16.7		635.7	26.5	63.0
		318.1			
Feb	15.0		636.2	26.5	56.6
		318.1			
Mar	21.9		636.6	26.5	82.6
		318.5			
Apr	26.5		637.9	26.6	99.6
		319.4			
May	31.1		639.9	26.7	116.5
		320.5			
Jun	31.6		643.9	26.8	117.9
		323.4			
Jul	32.5		647.2	27.0	120.4
		324.1			
Aug	33.3		647.9	27.0	123.3
		324.1			
Sep	31.5		646.8	27.0	116.7
		322.7			
Oct	32.3		647.0	27.0	119.6
		324.3			
Nov	27.6		651.2	27.1	101.8
		326.9			
Dec	23.4		654.5	27.3	85.7
		327.6			

(Table 9.13 continued)

TABLE 9.13 (CONTINUED)

Year and month (1)	Portland cement production (millions of barrels) (2)	12-month moving total (3)	2-month moving total of Col. 3 (4)	Recentered 12-month moving average (Col. 4 ÷ 24) (5)	Actual production as relative of moving average (Col. 2 ÷ Col. 5) (6)
1962 Jan	17.1	328.5	656.1	27.3	62.6
Feb	15.3	331.3	659.8	27.5	55.6
Mar	20.5	333.5	664.8	27.7	74.0
Apr	28.1	335.1	668.6	27.9	100.7
May	33.7	336.8	671.9	28.0	120.4
Jun	32.3	336.3	673.1	28.0	115.4
Jul	33.4				
Aug	36.1				
Sep	33.7				
Oct	33.9				
Nov	29.3				
Dec	22.9				

The Solution

Step 1 of the ratio-to-moving-average method outlined above appears as column 3 in Table 9.13. The first 12-month moving total, 248.9, is the sum of the 12 months of original data given in column 2. To derive the second 12-month moving total, 250.8, subtract the January 1952 value, 17.0, and add January 1953, 18.9. To derive the third 12-month moving total, 251.6, subtract the February 1952 value, 16.5, and add February 1953, 17.3. This process is continued until the last moving total is computed. Each 12-month moving total appears in column 3 between the middle two months of the 12 months to which it applies.

Step 2 appears in Table 9.13 as column 4. The first value, 499.7, is the sum of the first two items in column 3, 248.9 + 250.8. The 499.7 is on the same line as the original value for July 1952.

Step 3 in Table 9.13 appears as column 5. The first value, 20.8, is obtained by dividing 499.7 (from column 4) by 24, the number of months of original values included in 499.7.

Step 4 appears as column 6. The original value for July 1952, 21.3, is divided by 20.8, the recentered 12-month moving average (column 5), the result of which is multiplied by 100 in order to express it in percentage terms.

After column 6 is computed, the multiple frequency distribution table is set up (Table 9.14). The class limits for the relatives are established, and each entry in column 6 is properly entered on the table under the month and class interval indicated. This is step 5 of the ratio-to-moving-average method.

After all the entries from column 6 have been posted to the multiple frequency distribution table, the highest two values and the lowest two values are crossed out, and the sum of the middle six relatives is computed

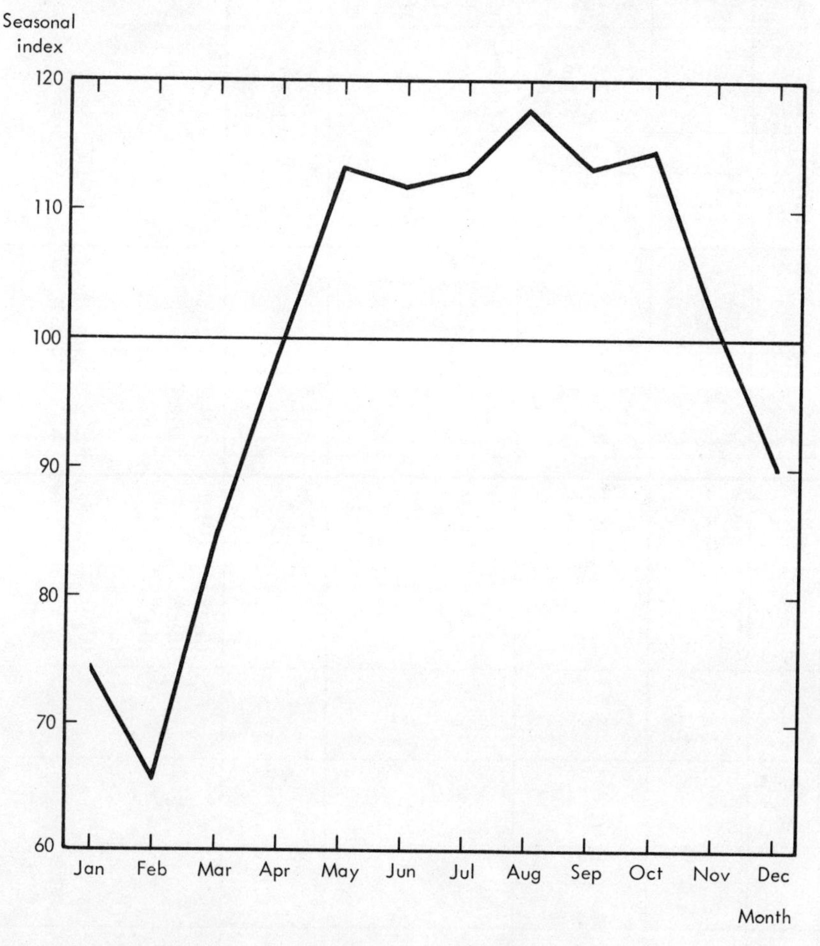

Fig. 9.2 Seasonal index of production of Portland cement in the United States, based on monthly data, 1952–1962
SOURCE: Table 9.15.

TABLE 9.14 PORTLAND CEMENT PRODUCTION (millions of barrels): MULTIPLE FREQUENCY TABLE FOR DERIVING A SEASONAL INDEX

Relatives	Jan	Feb	Mar	Apr	May	Jun	Jul	Aug	Sep	Oct	Nov	Dec
125–129.9								~~127.6~~ ~~126.6~~	~~127.7~~	~~125.4~~		
120–124.9					~~120.4~~		~~121.3~~ ~~120.8~~ 120.4	123.4 123.3 121.5		~~121.9~~		
115–119.9					~~118.7~~ 118.4 116.5	~~119.5~~ ~~118.8~~ 117.9 115.4		115.3	~~119.7~~ 117.8 116.8 116.7	119.6 118.0		
110–109.9					114.0 113.0 110.0	116.2	114.6 112.3 112.2 110.1 110.0	112.9 112.7 ~~112.1~~ ~~111.0~~	110.9 110.0	114.2 113.3 112.7 112.4 ~~111.6~~ ~~111.2~~		
105–109.9					109.8 ~~105.9~~	109.5 108.5 106.4			109.0 ~~108.7~~ ~~108.0~~			
100–104.9				~~102.5~~ ~~101.2~~ 100.7 100.7 100.0	~~104.0~~	~~103.2~~ ~~101.3~~	~~102.4~~				~~103.7~~ ~~103.2~~ 102.8 102.6 101.8 101.3 100.8 100.0	

95–99.9				99.6 99.1 97.3 ~~96.0~~							~~98.4~~ ~~95.3~~	~~97.2~~ ~~95.7~~
90–94.9			~~92.2~~ ~~91.8~~ 90.8 90.5 90.0	~~94.1~~								93.7 92.7 92.2 90.6
85–89.9	~~87.1~~		85.6									88.3 85.7 ~~85.5~~
80–84.9	~~84.2~~ 83.3 81.3		82.6				~~81.9~~					
75–79.9	76.9	~~79.0~~ ~~76.8~~ 75.7										~~77.4~~
70–74.9	73.1	72.7 72.1	74.0 ~~71.1~~									
65–69.9	69.0 66.7		~~68.4~~									
60–64.9	~~63.0~~ ~~62.6~~											
55–59.9		59.6 59.0 56.6 ~~55.7~~ ~~55.6~~										

SOURCE: Table 9.13.

and entered as indicated. This is step 6 of the ratio-to-moving-average method.

The sum of the middle six relatives for each month is then divided by 12 to obtain the unadjusted arithmetic mean for each month. This is step 7 of the ratio-to-moving-average method.

Finally, in step 8 of the ratio-to-moving-average method the twelve unadjusted means are added (Table 9.15). In this problem the total is

TABLE 9.15 PORTLAND CEMENT PRODUCTION, 1952–1962: COMPUTATION OF SEASONAL INDEX

Month	Sum of middle 6 relatives of actual production to 12-months moving average	Arithmetic mean	
		Unadjusted	Adjusted to equal 1200.0 (seasonal index)
Jan	450.3	75.1	74.8
Feb	395.7	66.0	65.8
Mar	513.5	85.6	85.3
Apr	597.4	99.6	99.2
May	681.7	113.6	113.2
Jun	673.9	112.3	111.9
Jul	679.6	113.3	112.9
Aug	709.1	118.2	117.8
Sep	681.2	113.5	113.1
Oct	690.2	115.0	114.6
Nov	609.3	101.6	101.2
Dec	543.2	90.5	90.2
	Total	1,204.3	1,200.0

SOURCE: Table 9.14.

1,204.3. Since the arithmetic means must be adjusted to equal 1,200, a correction factor is computed as follows:

$$\frac{1{,}200.0}{1{,}204.3} = 0.99643$$

Each of the unadjusted arithmetic means is multiplied by this correction factor, and this final result is the index of seasonal variation. The sum of the twelve values is 1,200, and the average is 100.0.

If the unadjusted total of arithmetic means were less than 1,200, a correction factor would be obtained in the same manner as follows:

$$\frac{1{,}200.0}{1{,}195.0} = 1.00418$$

Each unadjusted mean would be multiplied by 1.00418 to obtain the seasonal indexes.

9.13 CYCLICAL AND RANDOM FLUCTUATIONS

The two remaining components of the original data are the cyclical component (C) and the irregular or random component (I). The cyclical component reflects the impact of the business cycle, which has become an important area of research in economics since the second half of the nineteenth century. Before that time, alternating periods of prosperity and depression were associated with causes lying outside the area of economic science.

DEFINITIONS: The *cyclical component* (C) includes those changes which are attributable to business cycles or changing economic conditions.

The *irregular or random component* (I) includes variations due to strikes, floods, style changes, wars, or other unpredictable events.

PROCEDURES: The methods for measuring cyclical fluctuations follow. After measuring the cyclical component, the influence of the irregular or random component is still part of the cyclical values. To reduce the impact of the random component, the $C \times I$ residual is smoothed by means of a moving average.

Methods for Measuring Cyclical Fluctuations

The most intensive analysis of the American business cycle has been developed by the National Bureau of Economic Research in New York. The National Bureau analyzed over 800 monthly and quarterly time series in its study of cyclical movements in the United States. This sample of time series is representative of all major sectors of the economy and all phases of economic activity. As a result of this analysis, the National Bureau prepared a set of reference dates and durations of business cycles in the United States since 1854.

A less detailed and complex procedure used to analyze the cyclical-irregular component of a time series is known as the residual method, which follows.

9.14 MEASURING CYCLICAL FLUCTUATIONS BY THE RESIDUAL METHOD

The steps in this procedure as as follows:

1. Long-term growth is computed by means of the least squares method, and the trend line is derived. Thus, a trend value (T) may be computed for each period in the time series, usually each month.

2. The original data, or value for each month, symbol O or Y, is then divided by the growth line value (T), leaving the combined influences of $S \times C \times I$ as a remainder.

3. If the data of the time series are monthly or quarterly, the seasonal index is then computed (S).

4. The $S \times C \times I$ value is then divided by S to leave as a residual only the cyclical and irregular influences ($C \times I$).

5. Finally, this residual is generally smoothed out by means of a three- or five-month moving average to reduce as much as possible the impact of the irregular forces.

Although the complete analysis of a time series requires all the steps listed above, all of these steps are not always necessary. For example, if it is desired to compute only secular trend or only a seasonal pattern, then only the steps necessary for these computations are made. All the steps are required only when it is desired to analyze the cyclical pattern.

9.15 A PROBLEM IN TIME SERIES ANALYSIS: MEASUREMENT OF CYCLICAL FLUCTUATIONS

The Problem

Given a series of monthly data on Portland cement production in the United States for the years 1952 through 1962 (column 2 of Table 9.16) compute the cyclical influence.

The Solution

Step 1. In column 1, of Table 9.16 enter months and years as well as the X values, starting with zero as the first entry for the first month, and numbering all subsequent months consecutively. In column 2 enter each month's actual production.

Step 2. In column 3 of Table 9.16 enter each month's trend value. This is computed from the month-to-month trend line equation computed in the secular trend problem, question 6, page 200.

Step 3. In column 4 enter the seasonal index in ratio form for each month as computed in the problem on the seasonal index. This index is repeated for each year.

Step 4. Multiply each month's trend value by its corresponding seasonal index to obtain an expected value for each month based upon the influences of trend and seasonal forces. Enter these data in column 5.

Step 5. Divide the original, or actual, production value for each month (column 2) by the expected value for each month (column 5) to obtain the

TABLE 9.16 PORTLAND CEMENT PRODUCTION IN THE UNITED STATES, MONTHLY, JANUARY 1952–DECEMBER 1962: MEASUREMENT OF CYCLICAL FLUCTUATIONS BY THE RESIDUAL METHOD

Year, month, and X value[a] (1)		Portland cement production, Y (millions of barrels) (2)	Trend value,[b] T (3)	Seasonal index, S as ratio,[c] (4)	$(T \times S)$ (5)	$\frac{Y}{T \times S}$ equals $(C \times I)$ (6)	5-month moving total of Col. 6 (7)	5-month moving average or smoothed $C \times I$ (8)
1952 Jan	0	17.0	21.5	0.748	16.1	105.6		
Feb	1	16.5	21.5	0.658	14.1	117.0		
Mar	2	18.1	21.6	0.853	18.4	98.4	502.1	100.4
Apr	3	19.8	21.6	0.992	21.4	92.5	481.3	96.3
May	4	21.8	21.7	1.132	24.6	88.6	450.9	90.2
Jun	5	20.7	21.8	1.119	24.4	84.8	444.0	88.8
Jul	6	21.3	21.8	1.129	24.6	86.6	444.2	88.8
Aug	7	23.6	21.9	1.178	25.8	91.5	451.6	90.3
Sep	8	23.0	21.9	1.131	24.8	92.7	465.5	93.1
Oct	9	24.2	22.0	1.146	25.2	96.0	483.9	96.8
Nov	10	22.0	22.0	1.012	22.3	98.7	506.3	101.3
Dec	11	20.9	22.1	0.902	19.9	105.0	532.1	106.4
1953 Jan	12	18.9	22.2	0.748	16.6	113.9	542.4	108.5
Feb	13	17.3	22.2	0.658	14.6	118.5	542.3	108.5
Mar	14	20.2	22.3	0.853	19.0	106.3	529.0	105.8
Apr	15	21.8	22.3	0.992	22.1	98.6	505.5	101.1
May	16	23.3	22.4	1.132	25.4	91.7	481.9	96.4
Jun	17	22.7	22.4	1.119	25.1	90.4	467.3	93.5
Jul	18	24.1	22.5	1.129	25.4	94.9	461.7	92.3
Aug	19	24.3	22.5	1.178	26.5	91.7	465.0	93.0
Sep	20	23.8	22.6	1.131	25.6	93.0	472.4	94.5
Oct	21	24.7	22.7	1.146	26.0	95.0	475.6	95.1
Nov	22	22.5	22.7	1.012	23.0	97.8	488.0	97.6
Dec	23	20.2	22.8	0.902	20.6	98.1	506.9	101.4
1954 Jan	24	17.8	22.8	0.748	17.1	104.1	515.0	103.0
Feb	25	16.9	22.9	0.658	15.1	111.9	512.4	102.5
Mar	26	20.1	22.9	0.853	19.5	103.1	503.6	100.7
Apr	27	21.7	23.0	0.992	22.8	95.2	487.9	97.6
May	28	23.3	23.1	1.132	26.1	89.3	473.3	94.7
Jun	29	22.8	23.1	1.119	25.8	88.4	464.3	92.9
Jul	30	25.5	23.2	1.129	26.2	97.3	465.7	93.1
Aug	31	25.7	23.2	1.178	27.3	94.1	473.4	94.7
Sep	32	25.5	23.3	1.131	26.4	96.6	485.4	97.1
Oct	33	25.9	23.3	1.146	26.7	97.0	493.3	98.7
Nov	34	23.8	23.4	1.012	23.7	100.4	514.0	102.8
Dec	35	22.3	23.5	0.902	21.2	105.2	530.9	106.2
1955 Jan	36	20.2	23.5	0.748	17.6	114.8	544.8	109.0
Feb	37	17.6	23.6	0.658	15.5	113.5	549.9	110.0
Mar	38	22.3	23.6	0.853	20.1	110.9	545.4	109.1
Apr	39	24.8	23.7	0.992	23.5	105.5	531.4	106.3
May	40	27.0	23.7	1.132	26.8	100.7	519.4	103.9
Jun	41	26.8	23.8	1.119	26.6	100.8	507.4	101.5
Jul	42	27.3	23.8	1.129	26.9	101.5	501.5	100.3
Aug	43	27.9	23.9	1.178	28.2	98.9	502.3	100.5
Sep	44	27.0	24.0	1.131	27.1	99.6	503.5	100.7
Oct	45	27.9	24.0	1.146	27.5	101.5	508.5	101.7

(Table 9.16 continued)

TABLE 9.16 (CONTINUED)

Year, month, and X value[a] (1)		Portland cement production, Y (millions of barrels) (2)	Trend value,[b] T (3)	Seasonal index, S as ratio,[c] (4)	$(T \times S)$ (5)	$\frac{Y}{T \times S}$ equals $(C \times I)$ (6)	5-month moving total of Col. 6 (7)	5-month moving average or smoothed $C \times I$ (8)
1955 Nov	46	24.9	24.1	1.012	24.4	102.0	527.8	105.6
Dec	47	23.1	24.1	0.902	21.7	106.5	551.5	110.3
1956 Jan	48	21.4	24.2	0.748	18.1	118.2	563.0	112.6
Feb	49	19.6	24.2	0.658	15.9	123.3	568.9	113.8
Mar	50	23.4	24.3	0.853	20.7	113.0	569.6	113.9
Apr	51	26.1	24.4	0.992	24.2	107.9	556.5	111.3
May	52	29.6	24.4	1.132	27.6	107.2	539.7	107.9
Jun	53	28.8	24.5	1.119	27.4	105.1	530.5	106.1
Jul	54	29.5	24.5	1.129	27.7	106.5	525.5	105.1
Aug	55	30.1	24.6	1.178	29.0	103.8	521.1	104.2
Sep	56	28.6	24.6	1.131	27.8	102.9	519.2	103.8
Oct	57	29.1	24.7	1.146	28.3	102.8	521.6	104.3
Nov	58	25.9	24.8	1.012	25.1	103.2	521.6	104.3
Dec	59	24.4	24.8	0.902	22.4	108.9	527.2	105.4
1957 Jan	60	19.3	24.9	0.748	18.6	103.8	530.5	106.1
Feb	61	17.8	24.9	0.658	16.4	108.5	524.1	104.8
Mar	62	22.6	25.0	0.853	21.3	106.1	512.0	102.4
Apr	63	24.0	25.0	0.992	24.8	96.8	502.5	100.5
May	64	27.5	25.1	1.132	28.4	96.8	465.2	93.0
Jun	65	26.5	25.1	1.119	28.1	94.3	464.5	92.9
Jul	66	20.3	25.2	1.129	28.5	71.2	475.7	95.1
Aug	67	31.4	25.3	1.178	29.8	105.4	482.3	96.5
Sep	68	30.9	25.3	1.131	28.6	108.0	485.3	97.1
Oct	69	30.1	25.4	1.146	29.1	103.4	511.5	102.3
Nov	70	25.0	25.4	1.012	25.7	97.3	501.4	100.3
Dec	71	22.4	25.5	0.902	23.0	97.4	477.3	95.5
1958 Jan	72	18.2	25.5	0.748	19.1	95.3	456.1	91.2
Feb	73	14.1	25.6	0.658	16.8	83.9	452.9	90.6
Mar	74	18.0	25.7	0.853	21.9	82.2	455.8	91.2
Apr	75	24.0	25.7	0.992	25.5	94.1	464.7	92.9
May	76	29.3	25.8	1.132	29.2	100.3	482.8	96.6
Jun	77	30.1	25.8	1.119	28.9	104.2	504.5	100.9
Jul	78	29.8	25.9	1.129	29.2	102.0	517.9	103.6
Aug	79	31.7	25.9	1.178	30.5	103.9	527.3	105.5
Sep	80	31.6	26.0	1.131	29.4	107.5	529.2	105.8
Oct	81	32.8	26.1	1.146	29.9	109.7	527.2	105.4
Nov	82	28.0	26.1	1.012	26.4	106.1	518.2	103.6
Dec	83	23.6	26.2	0.902	23.6	100.0	507.2	101.4
1959 Jan	84	18.6	26.2	0.748	19.6	94.9	506.0	101.2
Feb	85	16.7	26.3	0.658	17.3	96.5	511.0	102.2
Mar	86	24.3	26.3	0.853	22.4	108.5	522.7	104.5
Apr	87	29.1	26.4	0.992	26.2	111.1	540.6	108.1
May	88	33.4	26.4	1.132	29.9	111.7	558.1	111.6
Jun	89	33.5	26.5	1.119	29.7	112.8	560.8	112.2
Jul	90	34.2	26.6	1.129	30.0	114.0	557.6	111.5
Aug	91	34.8	26.6	1.178	31.3	111.2	547.5	109.5

(Table 9.16 continued)

TABLE 9.16 (CONTINUED)

Year, month, and X value[a] (1)	Portland cement production, Y (millions of barrels) (2)	Trend value, T[b] (3)	Seasonal index, S as ratio,[c] (4)	(T × S) (5)	$\frac{Y}{T \times S}$ equals (C × I) (6)	5-month moving total of Col. 6 (7)	5-month moving average or smoothed C × I (8)
1959 Sep 92	32.6	26.7	1.131	30.2	107.9	531.0	106.2
Oct 93	31.1	26.7	1.146	30.6	101.6	516.6	103.3
Nov 94	26.1	26.8	1.012	27.1	96.3	498.4	99.7
Dec 95	24.1	26.8	0.902	24.2	99.6	480.9	96.2
1960 Jan 96	18.7	26.9	0.748	20.1	93.0	459.3	91.9
Feb 97	16.1	27.0	0.658	17.8	90.4	463.4	92.7
Mar 98	18.4	27.0	0.853	23.0	80.0	467.4	93.5
Apr 99	27.0	27.1	0.992	26.9	100.4	479.3	95.9
May 100	31.8	27.1	1.132	30.7	103.6	493.1	98.6
Jun 101	31.9	27.2	1.119	30.4	104.9	516.5	103.3
Jul 102	32.0	27.2	1.129	30.7	104.2	516.4	103.3
Aug 103	33.3	27.3	1.178	32.2	103.4	512.8	102.6
Sep 104	31.1	27.4	1.131	31.0	100.3	503.2	100.6
Oct 105	31.4	27.4	1.146	31.4	100.0	481.7	96.3
Nov 106	26.5	27.5	1.012	27.8	95.3	459.4	91.9
Dec 107	20.5	27.5	0.902	24.8	82.7	441.5	88.3
1961 Jan 108	16.7	27.6	0.748	20.6	81.1	434.3	86.9
Feb 109	15.0	27.6	0.658	18.2	82.4	435.4	87.1
Mar 110	21.9	27.7	0.853	23.6	92.8	451.4	90.3
Apr 111	26.5	27.7	0.992	27.5	96.4	471.6	94.3
May 112	31.1	27.8	1.132	31.5	98.7	492.4	98.5
Jun 113	31.6	27.9	1.119	31.2	101.3	500.5	100.1
Jul 114	32.5	27.9	1.129	31.5	103.2	503.2	100.7
Aug 115	33.3	28.0	1.178	33.0	100.9	505.1	101.0
Sep 116	31.5	28.0	1.131	31.7	99.4	501.0	100.2
Oct 117	32.3	28.1	1.146	32.2	100.3	489.9	98.0
Nov 118	27.6	28.1	1.012	28.4	97.2	469.7	93.9
Dec 119	23.4	28.2	0.902	25.4	92.1	452.6	90.5
1962 Jan 120	17.1	28.3	0.748	21.2	80.7	437.0	87.4
Feb 121	15.3	28.3	0.658	18.6	82.3	439.4	87.9
Mar 122	20.5	28.4	0.853	24.2	84.7	451.6	90.3
Apr 123	28.1	28.4	0.992	28.2	99.6	472.2	94.4
May 124	33.7	28.5	1.132	32.3	104.3	493.3	98.7
Jun 125	32.3	28.5	1.119	31.9	101.3	515.4	103.1
Jul 126	33.4	28.6	1.129	32.3	103.4	519.5	103.9
Aug 127	36.1	28.7	1.178	33.8	106.8	517.9	103.6
Sep 128	33.7	28.7	1.131	32.5	103.7	517.3	103.5
Oct 129	33.9	28.8	1.146	33.0	102.7	501.6	100.3
Nov 130	29.3	28.8	1.012	29.1	100.7		
Dec 131	22.9	28.9	0.902	26.1	87.7		

[a] The X value is the value which appears in the trend equation. With the origin as January 15, 1952, and X units equal to one month, it is a consecutive count of the months from 0 to 131.

[b] $Y_c = 21.4750 + 0.0565X$; origin: January 15, 1952, and X units equal to one month. Each Y_c value is rounded to one decimal.

[c] Computed as mean of middle 6 ratios of actual values to recentered 12-month moving averages for years 1952–1962. This index was derived in the problem on the seasonal index.

TABLE 9.17 CONSUMPTION OF CIGARETTES IN THE UNITED STATES, BASED UPON PRODUCERS' PAYMENTS OF FEDERAL TAXES, 1954–1964

(billions of dollars)

	1954	1955	1956	1957	1958	1959	1960	1961	1962	1963	1964
Jan	28.9	30.4	32.9	36.0	35.8	36.2	37.6	38.9	41.1	43.5	41.0
Feb	26.7	28.7	30.7	31.7	31.4	34.6	35.2	37.4	35.8	38.0	29.2
Mar	32.3	33.7	32.5	33.2	33.0	35.5	40.3	42.4	42.6	39.6	37.9
Apr	30.5	28.8	30.2	32.1	35.7	38.1	36.9	37.2	38.6	42.3	43.7
May	31.9	34.5	36.2	38.2	37.6	37.3	41.4	44.4	45.1	48.2	41.7
Jun	35.0	35.6	34.3	34.2	38.6	38.4	43.6	44.0	41.3	41.6	45.2
Jul	29.0	28.6	31.0	35.2	36.8	39.9	35.7	35.9	39.4	42.4	42.6
Aug	34.6	36.8	37.6	38.0	39.6	40.9	44.6	47.2	47.3	47.0	44.4
Sep	32.0	32.1	30.4	34.7	38.1	39.2	40.9	39.6	40.5	42.4	43.3
Oct	31.6	32.9	37.2	38.2	40.9	43.1	39.8	45.4	45.6	46.7	47.1
Nov	29.7	32.6	33.6	31.5	34.8	36.2	40.3	42.6	42.5	41.3	41.5
Dec	26.7	27.4	25.1	26.4	34.0	34.3	33.8	33.3	34.7	36.7	39.9
Total	368.9	382.1	391.7	409.4	436.3	453.7	470.1	488.3	494.5	509.7	497.5
Monthly average	30.7	31.8	32.6	34.1	36.4	37.8	39.2	40.7	41.2	42.5	41.5

SOURCE: U.S. Department of Commerce, Office of Business Economics, *Survey of Current Business*, p. S-30.

residual influence of the cyclical and irregular forces ($C \times I$). Enter these data in column 6.

Step 6. By means of a five-month moving average, smooth out the influence of irregular forces in the column 6 data in order to obtain a smoother representation of the cyclical influence. Enter five-month moving totals in column 7 and the five-month moving averages in column 8.

EXERCISES

1. Define or explain the following:

Annual trend equation	Origin year
Calendar variation	Original data
Constant dollars	Population adjustment
Cyclical movement	Price adjustment
Decompose a series	Random forces
Deflation	Ratio-to-moving-average method
Eliminate the influence of a component	Real dollars
	Seasonal index
Graphic method for determining trend	Seasonal variation
	Secular trend
Irregular forces	Shifting the origin
Long-term forecast	Short-term forecast
Month-to-month trend equation	Time series
	Time series analysis
Moving average method	Transformation in the trend line
Multiple frequency table	Trend equation

2. Table 9.17 provides data on cigarette consumption by month as well as annual totals and monthly averages.

    a. Using the monthly averages, with 1954 as the origin year, solve the normal equations to obtain a and b. Set up the linear equation of trend, citing the origin and X units.
    b. Compute Y_c for any two X values.
    c. Plot the original data and the trend equation on a sheet of arithmetic graph paper.
    d. Using the monthly averages, with 1959 as the origin year, solve the normal equations to obtain a and b. Set up the linear equation of trend, citing the origin and X units.
    e. Is the trend line derived in answer to question (d) above the same as that derived in answer to question (a) above? Explain.

f. Transform the linear equation of trend derived in answer to question (d) above from X units = one year to X units = one month. Cite the origin and X units.
g. Using the equation computed in answer to question (f) above, shift the origin from June 30/July 1, 1959, to January 15, 1954.
h. Using the annual trend equation with origin at June 30/July 1, 1959, compute the trend value for 1969.
i. Using the month-to-month trend equation derived in answer to question (g) above, compute the trend value for October 1968.

3. Using the annual totals data of Table 9.17:

a. With 1959 as the origin year, solve the normal equations to obtain a and b. Set up the linear equation of trend, citing the origin and X units.
b. Compute Y_c for any two values of X.
c. Plot the original data and the trend equation on a sheet of arithmetic graph paper.

4. Answer the following:

a. List the steps required to compute a seasonal index by the ratio-to-moving-average method.
b. Using the monthly data of Table 9.17 (132 entries), compute the seasonal index for cigarette consumption by the ratio-to-moving average method.
c. Plot the seasonal index computed above on a sheet of arithmetic graph paper.

5. Given the monthly data in Table 9.17,

a. Compute the cyclical influence.
b. By means of a five-month moving average, smooth out the influence of irregular forces.

6. Using the data in Table 9.18, compute a seasonal index by the ratio-to-moving-average method.

7. Given the data in Table 9.19:

a. By the method of least squares, compute the linear equation of trend.
b. Assuming a continuation of this trend, electric power sales to ultimate consumers for 1968 should be ______.
c. The trend value for 1960 is ______.

TABLE 9.18 VALUE OF NEW RESIDENTIAL CONSTRUCTION PUT IN PLACE IN THE UNITED STATES, 1956–1964[a]
(millions of dollars)

Year	Total	Quarter			
		First	Second	Third	Fourth
1956	$20,178	$4,154	$5,255	$5,663	$5,106
1957	19,006	3,880	4,808	5,350	4,968
1958	19,789	3,786	4,673	5,587	5,743
1959	24,251[b]	4,775	6,251	7,018	6,207
1960	21,706	4,609	5,851	5,952	5,294
1961	21,680	4,058	5,639	6,161	5,822
1962	24,292	4,516	6,424	7,064	6,288
1963	25,843	4,823	6,767	7,335	6,918
1964	26,560	5,347	7,106	7,469	6,638

[a] Includes new housing units, additions and alterations, and nonhousekeeping buildings. Excludes farm construction.

[b] Data for 1959 and forward include Alaska and Hawaii.

SOURCE: 1956–1963 U.S. Department of Commerce, Bureau of the Census, *Value of New Construction Put in Place*, 1946–1963 *Revised*. Construction Report C30–61 Supplement, 1964. 1964 data: U.S. Department of Commerce, Office of Business Economics, *Survey of Current Business*, March 1965, p. S-9.

TABLE 9.19 ELECTRIC POWER SALES TO ULTIMATE CONSUMERS, 1957–1963
(billions of kilowatt hours)

Year	Electric power sales
1957	5.7
1958	6.2
1959	6.8
1960	7.6
1961	8.4
1962	9.2
1963	10.1

10

Index Numbers

Index numbers are statistical devices used by businessmen, government officials, economists, and others to obtain an understandable view of over-all changes in prices, volume of production, and values. Actual figures are often too voluminous or diverse to comprehend. An index number, however, is a summary figure which is typical or representative of a large number of actual values. In this sense an index number relates to the concept of the average. Index numbers are most useful in facilitating comparisons over time. Thus, certain aspects of time series analysis play a role in the construction and use of index numbers.

In this chapter, the term index number *designates a single ratio in percentage form used to measure the average or combined change of variables between two different times. While the chapter concentrates on economic variables over time, the concept of index numbers is applicable to noneconomic data, for example, measurement of the efficiency of teachers.*

The first use of the technique of index numbers appears in the literature in 1764 *when Carli* (1720–1795), *an Italian analyst, measured the impact of the importation of silver from America upon the prices of wine, oil, and grain in Italy. He averaged price relatives for these commodities for the years* 1500 *and* 1750 *to measure changes in purchasing power.*

Historically, economists and statisticians have been interested primarily in the measurement of changes in the price level. By means of the Consumer Price Index and the Wholesale Price Index (discussed below, in Sections 10.7 *and* 10.8) *much has been accomplished toward the achievement of this goal. While ultimately economists would like to develop a cost-of-living index, this has not yet been accomplished.*

This chapter defines an index number, classifies index numbers, and analyzes the major types of index number procedures. It illustrates the technique of weighting as used in index number construction, including the formulas for weighted aggregative index numbers.

Special index number techniques, such as splicing and changing the base year, are explained. The chapter then considers the basic decisions which must be made in constructing an index number series.

The characteristics, sources, uses, and limitations of the Consumer Price Index and the Wholesale Price Index are analyzed, and the relationship between the two indexes is summarized.

10.1 INTRODUCTION TO INDEX NUMBERS

DEFINITION: An *index number* is a statistical measure of fluctuations in a variable, arranged in the form of a series, and using a base period for making comparisons.

Index numbers may be *classified* into four groups:

1. Price indexes. Examples of price indexes are as follows:

   a. The Consumer Price Index, prepared monthly by the U.S. Department of Labor, Bureau of Labor Statistics.
   b. The Wholesale Price Index, prepared monthly by the U.S. Department of Labor, Bureau of Labor statistics. A weekly and a daily index are also prepared by the same agency.
   c. Farm price indexes. Two series are available.
      (1) Prices farmers pay for the goods they buy.
      (2) Prices farmers receive for their production.
      These are prepared monthly by the U.S. Department of Agriculture, Statistical Reporting Service.
   d. Stock price indexes. The *New York Times* publishes a daily series on stock prices.

2. Indexes of physical quantity: The outstanding example is the monthly Index of Industrial Production, prepared by the Board of Governors of the Federal Reserve System. The summary indexes include total production, minerals production, durable manufactures, and nondurable manufactures.

3. Value indexes: An example of a value index is the Federal Reserve Board Index of Department Store Sales, issued monthly.

4. Special-purpose index numbers: Some outstanding examples include railroad freight carloadings, number of persons employed, average annual expenditures per pupil, and others.

10.2 MAJOR TYPES OF INDEX NUMBER PROCEDURES

Procedures for constructing index numbers may be classified into two groups: *aggregative* and *averages of relatives*. Within each of these groups, index numbers may be classified as either simple indexes or weighted indexes.

Aggregative Indexes

SIMPLE: This type of index series is known as an aggregate of actual prices.

WEIGHTED AGGREGATIVE INDEX (four types)

1. Base-year quantity weights are used (Laspeyres' formula).
2. Given-year quantity weights are used (Paasche's (1851–1925) formula).

3. Uses a combination of base-year and given-year quantity weights. (The Marshall [1842–1924]-Edgeworth [1845–1926] formula.)

4. The square root of the product of the indexes derived from the Laspeyres and Paasche formulas, known as Fisher's (1867–1947) "ideal" index, is used.

Averages of Relatives

SIMPLE: This type is derived by computing an arithmetic average of relative prices.

WEIGHTED ARITHMETIC AVERAGE OF RELATIVE PRICES: Value weights are used, consisting of the product of the price and quantity for each item included.

10.3 PROBLEMS ILLUSTRATING MAJOR TYPES OF INDEX NUMBER PROCEDURES

Given the data in Table 10.1.

1. Construct a simple arithmetic average of price relatives, using 1956 = 100.0 as the base year.

TABLE 10.1 AVERAGE ANNUAL PRICES FOR COMMODITIES A, B, AND C, 1956–1962

Year	Prices in cents for		
	Commodity A	Commodity B	Commodity C
1956	22.8	14.7	31.7
1957	23.8	14.4	30.5
1958	30.4	15.0	32.9
1959	39.1	16.1	37.6
1960	40.9	16.8	39.3
1961	41.2	17.3	40.4
1962	41.8	17.8	42.7

2. Construct a simple aggregate of actual prices, using 1956 = 100.0 as the base year.
3. Construct a weighted average of price relatives, using a value weight derived for the base year, using 1956 = 100.0 as the base year.
4. Construct a weighted aggregative index, weighted by base-year quantities, using 1956 = 100.0 as the base year.
5. Compare the weighted aggregative index computed in (4) above with the weighted average of price relatives computed in (3) above.

QUESTION 1: Construct a simple arithmetic average of price relatives.

FORMULA:

$$P_n = \frac{\Sigma (p_n/p_o)}{n}$$

where P = price index number
p = price for individual commodity
$_n$ = given period
$_o$ = base period
N = number of items
Σ = sum of

STEPS

1. For commodity A:

$$P_{1957} = \frac{p_{1957}}{p_{1956}} = \frac{23.8}{22.8} \times 100 = 104.4$$

The prices for commodity A are given. The same computation is repeated for commodities B and C for the year 1957.

2. To obtain the simple arithmetic average of price relatives for 1957, the price relatives for commodities A, B, and C, derived in step 1 above, are added and divided by 3, as follows:

$$P_{1957} = \frac{104.4 + 98.0 + 96.2}{3} = \frac{298.6}{3} = 99.5$$

3. Steps 1 and 2 are repeated for each succeeding year. The denominator of the price relative for each commodity is the price of that commodity in 1956, which is the base year. For example, the price relative for 1958 for commodity A is

$$\frac{p_{1958}}{p_{1956}} = \frac{30.4}{22.8} \times 100 = 133.3$$

The results are summarized in Table 10.2.

TABLE 10.2 PRICE RELATIVES FOR COMMODITIES A, B, AND C, 1956–1962
(1956 = 100.0)

Year	Price relatives			Simple arithmetic average of price relatives
	Commodity A	Commodity B	Commodity C	
1956	100.0	100.0	100.0	100.0
1957	104.4	98.0	96.2	99.5
1958	133.3	102.0	103.8	113.0
1959	171.5	109.5	118.6	133.2
1960	179.4	114.3	124.0	139.2
1961	180.7	117.7	127.4	141.9
1962	183.3	121.1	134.7	146.4

QUESTION 2: Construct a simple aggregate of actual prices, using 1956 = 100.0 as the base year.

FORMULA:

$$P_n = \frac{\Sigma p_n}{\Sigma p_o}$$

STEPS

1. $P_{1957} = \frac{\Sigma p_{1957}}{\Sigma p_{1956}} = \frac{68.7}{69.2} \times 100 = 99.3$ = price index for 1957.

2. To derive the price index for 1958, the sum of 1958 prices (78.3 cents) is divided by the sum of 1956 prices (69.2 cents). Because 1956 is the base year, the sum of 1956 prices is used each year as the denominator of the

fraction. The same procedure is used for each succeeding year. The results are shown in Table 10.3.

TABLE 10.3 AGGREGATE OF PRICES FOR COMMODITIES A, B, AND C, 1956–1962

Year	Prices in cents for				Simple index of aggregate prices (1956 = 100.0)
	Commodity A	Commodity B	Commodity C	Total (A + B + C)	
1956	22.8	14.7	31.7	69.2	100.0
1957	23.8	14.4	30.5	68.7	99.3
1958	30.4	15.0	32.9	78.3	113.2
1959	39.1	16.1	37.6	92.8	134.1
1960	40.9	16.8	39.3	97.0	140.2
1961	41.2	17.3	40.4	98.9	142.9
1962	41.8	17.8	42.7	102.3	147.8

QUESTION 3. Construct a weighted average of price relatives, using a value weight derived for the base year (1956 = 100.0).

Derivation of the value weight for the base year is shown in Table 10.4.

TABLE 10.4 VALUE OF EXPENDITURES ON COMMODITIES A, B, AND C IN BASE YEAR 1956

Commodity	Year 1956		
	Quantity purchased (thousands of units)	Price per unit (cents)	Value weight ($p_o q_o$)
A	9	22.8	\$2.052
B	7	14.7	1.029
C	6	31.7	1.902
Total			\$4.983

Derivation of the weighted price relatives is shown in Table 10.5.

FORMULA:

$$P_n = \frac{\sum \frac{p_n}{p_o}(p_o q_o)}{\Sigma\, p_o q_o}$$

TABLE 10.5 WEIGHTED PRICE RELATIVES OF COMMODITIES A, B, AND C, 1957–1962

Commodity	Value weight[a] ($p_o q_o$)	Year 1957		Year 1958		Year 1959	
		Price relative	Weight × price relative	Price relative	Weight × price relative	Price relative	Weight × price relative
A	$2.052	104.4	214.23	133.3	273.53	171.5	351.92
B	1.029	98.0	100.84	102.0	104.96	109.5	112.68
C	1.902	96.2	182.97	103.8	197.43	118.6	225.58
Total	$4.983		498.04		575.92		690.18

Commodity	Value weight[a] ($p_o q_o$)	Year 1960		Year 1961		Year 1962	
		Price relative	Weight × price relative	Price relative	Weight × price relative	Price relative	Weight × price relative
A	$2.052	179.4	368.13	180.7	370.80	183.3	376.13
B	1.029	114.3	117.61	117.7	121.11	121.1	124.61
C	1.902	124.0	235.85	127.4	242.31	134.7	256.20
Total	$4.983		721.59		734.22		756.94

[a] From Table 10.4.

STEPS

1. To obtain the price index for 1957, substitute values from Table 10.5 in the formula above. Thus, the price index for 1957 is:

$$\frac{\sum \frac{p_{1957}}{p_{1956}} (p_{1956} q_{1956})}{\sum p_{1956} q_{1956}} = \frac{498.04}{4.983} = 99.9$$

2. For 1958: 575.92/4.983 = 115.6
3. For 1959: 690.18/4.983 = 138.5
4. For 1960: 721.59/4.983 = 144.8
5. For 1961: 734.22/4.983 = 147.3
6. For 1962: 756.94/4.983 = 151.9

The index number for the base year (1956) is 498.3/4.983 = 100.0.

QUESTION 4. Construct a weighted aggregative index, weighted by base-year quantities, using 1956 = 100.0 as the base year.

Derivation of weighted prices for each year is indicated in Table 10.6.

TABLE 10.6 DERIVATION OF WEIGHTED PRICES FOR COMMODITIES A, B, AND C, 1956–1962

Commodity	Quantity purchased in 1956 (thousands of units) (q_{56})	Year 1956 Price (p_{56})	Year 1956 Weighted price ($p_{56}q_{56}$)	Year 1957 Price (p_{57})	Year 1957 Weighted price ($p_{57}q_{56}$)
A	9	$0.228	$2.052	$0.238	$2.142
B	7	0.147	1.029	0.144	1.008
C	6	0.317	1.902	0.305	1.830
Total			$4.983		$4.980

Year 1958 Price (p_{58})	Year 1958 Weighted price ($p_{58}q_{56}$)	Year 1959 Price (p_{59})	Year 1959 Weighted price ($p_{59}q_{56}$)	Year 1960 Price (p_{60})	Year 1960 Weighted price ($p_{60}q_{56}$)
$0.304	$2.736	$0.391	$3.519	$0.409	$3.681
0.150	1.050	0.161	1.127	0.168	1.176
0.329	1.974	0.376	2.256	0.393	2.358
	$5.760		$6.902		$7.215

Year 1961 Price (p_{61})	Year 1961 Weighted price ($p_{61}q_{56}$)	Year 1962 Price (p_{62})	Year 1962 Weighted price ($p_{62}q_{56}$)
$0.412	$3.708	$0.418	$3.762
0.173	1.211	0.178	1.246
0.404	2.424	0.427	2.562
	$7.343		$7.570

FORMULA:

$$P_n = \frac{\Sigma\, p_n q_o}{\Sigma\, p_o q_o}$$

This is the Laspeyres' formula. Prices are weighted by base-year quantities.

STEPS

1. To obtain the price index for 1957, substitute values from Table 10.6 in the formula above. Thus, the price index for 1957 is

$$\frac{\Sigma\, p_{57}q_{56}}{\Sigma\, p_{56}q_{56}} = \frac{\$4.980}{\$4.983} \times 100 = 99.9$$

2. For 1958: \$5.760/\$4.983 × 100 = 115.6
3. For 1959: \$6.902/\$4.983 × 100 = 138.5
4. For 1960: \$7.215/\$4.983 × 100 = 144.8
5. For 1961: \$7.343/\$4.983 × 100 = 147.4
6. For 1962: \$7.570/\$4.983 × 100 = 151.9

The index number for the base year (1956) is \$4.983/\$4.983 × 100 = 100.0.

QUESTION 5. Compare the weighted aggregative index and the weighted arithmetic average of price relatives. What is the conclusion to be drawn?

Table 10.7 presents the results of the computations for the weighted indexes as indicated in answer to questions 3 and 4 above.

TABLE 10.7 COMPARISON OF WEIGHTED AGGREGATIVE INDEX WITH WEIGHTED ARITHMETIC AVERAGE OF PRICE RELATIVES, 1956–1962

Year	Weighted aggregative index	Weighted arithmetic average of price relatives
1956	100.0	100.0
1957	99.9	99.9
1958	115.6	115.6
1959	138.5	138.5
1960	144.8	144.8
1961	147.4[a]	147.3[a]
1962	151.9	151.9

[a] The difference of 0.1 is due to rounding.

Conclusion: The weighted aggregative index and the weighted arithmetic average of price relatives yield the same results. The reason for this equality may be indicated by the following relationship:

The formula for the weighted aggregative index	equals	The formula for the weighted arithmetic average of price relatives
$\frac{\Sigma p_n q_o}{\Sigma p_o q_o}$	$=$	$\frac{\sum \frac{p_n}{p_o}(p_o q_o)}{p_o q_o}$

Therefore,

$$\frac{\Sigma p_n q_o}{\Sigma p_o q_o} = \frac{\Sigma p_n q_o}{\Sigma p_o q_o}$$

10.4 WEIGHTED AGGREGATIVE INDEX NUMBERS: SUMMARY OF FORMULAS

The formulas for constructing weighted aggregative index numbers are:

1. *Laspeyres' formula* uses base-year quantity weights. The application of this formula was illustrated in question 4 of the previous problem.

$$P_n = \frac{\Sigma p_n q_o}{\Sigma p_o q_o}$$

2. *Paasche formula* uses given-year quantity weights. In this formula the weights are changed with each successive year. This formula was developed to reflect changes in the relative importance of items in successive periods. The problem in attempting to apply this formula is that quantity weights are not readily available for current periods and are generally difficult to develop.

$$P_n = \frac{\Sigma p_n q_n}{\Sigma p_o q_n}$$

Illustration: If 1956 = 100.0, then indexes for 1957, 1958, and 1959 are as follows:

$$P_{57} = \frac{\Sigma p_{57} q_{57}}{\Sigma p_{56} q_{57}}$$

$$P_{58} = \frac{\Sigma p_{58} q_{58}}{\Sigma p_{56} q_{58}}$$

$$P_{59} = \frac{\Sigma p_{59} q_{59}}{\Sigma p_{56} q_{59}}$$

3. *Marshall-Edgeworth formula* uses a combination of base-year and given-year weights. The combination of weights utilizes the advantages which accrue from both base-year and current-year quantities. The problem, as in the Paasche formula, is in obtaining current-year quantity weights for the large number of items included in an index.

$$P_n = \frac{\Sigma\, p_n(q_o + q_n)}{\Sigma\, p_o(q_o + q_n)}$$

4. *Fisher's "ideal" index*. This index number formula, developed by Irving Fisher, is called *ideal* because it meets certain index number tests which may be applied to index numbers. The formula is a geometric average of the Laspeyres and Paasche formulas, each of which has a bias in opposite directions.

$$P_n = \sqrt{\left(\frac{\Sigma\, p_n q_o}{\Sigma\, p_o q_o}\right)\left(\frac{\Sigma\, p_n q_n}{\Sigma\, p_o q_n}\right)}$$

The formulas listed above have been modified in a number of ways to meet specific needs.

10.5 SPECIAL INDEX NUMBER PROCEDURES

Two special procedures are used in working with index numbers: *splicing* and *changing the base year*. When an index is revised by introducing new commodities into it or by changing its weights, it may no longer be exactly comparable with the old series. In order to make the series continuous, the old indexes are spliced with the new indexes.

DEFINITION: *Splicing* is the statistical procedure which connects an old index number series with a revised series in order to make the series continuous.

After a period of years, it is often desired to change the base year, or years, of an index number series to a more recent year, or years. This is generally done to make the series more meaningful for present users. It is important to note that *changing the base year* in no way changes the relationship of the individual index number values to each other. The existing value relationship is retained even though the base year, or years, may be changed.

DEFINITION: *Changing the base year* of an index number series is a statistical procedure which shifts the base year, or years, to another year, or years.

These procedures are illustrated in the following examples.

Example 1: Splicing Index Numbers

In 1958 the ABC Manufacturing Company changed the weights and revised its index. Its former index was on a 1950 = 100.0 base, and the base for the revised index is 1958 = 100.0.

PROBLEM: Given the data in Table 10.8, splice the two series.

TABLE 10.8 ILLUSTRATION OF SPLICING OF INDEX NUMBERS

Year	1950 = 100.0 base	1958 = 100.0 base
1950	100.0	
1951	104.7	
1952	106.3	
1953	109.1	
1954	107.5	
1955	110.8	
1956	113.6	
1957	116.5	
1958	119.2	100.0
1959		101.5
1960		103.2
1961		104.1
1962		105.9

Solution: To splice one index number series with another, it is necessary to have the two series overlap for one year. In this example, the year 1958 has a value on the old and the new base. Each index number on the old base is divided by 119.2. Thus, the index number on the new base (1958 = 100.0) for 1950 equals 100.0 ÷ 119.2; for 1951, 104.7 ÷ 119.2; for 1952, 106.3 ÷ 119.2, and so forth. The spliced index is shown in Table 10.9.

Example 2: Changing the Base Year

Assuming that the ABC Manufacturing Company had not changed its index weights, and continued to construct its index number series on the 1950 = 100.0 base, then the series in Table 10.10 would have evolved.

TABLE 10.9 ILLUSTRATION OF A SPLICED INDEX NUMBER SERIES

Year	Base	
	1950 = 100.0	1958 = 100.0
1950	100.0	83.9
1951	104.7	87.8
1952	106.3	89.2
1953	109.1	91.5
1954	107.5	90.2
1955	110.8	93.0
1956	113.6	95.3
1957	116.5	97.7
1958	119.2	100.0
1959		101.5
1960		103.2
1961		104.1
1962		105.9

TABLE 10.10 INDEX NUMBERS FOR ABC MANUFACTURING COMPANY (1950 = 100.0)

Year	1950 = 100.0 base
1950	100.0
1951	104.7
1952	106.3
1953	109.1
1954	107.5
1955	110.8
1956	113.6
1957	116.5
1958	119.2
1959	120.4
1960	124.9
1961	127.6
1962	128.3

PROBLEM: Change the base from 1950 = 100.0 to 1955 = 100.0.

Solution: To change the base year, each index number in the series is divided by the index number of the new base year. Thus, each index number must be divided by 110.8, the index for 1955. For 1950, 100.0 ÷ 110.8 = 90.3; for 1951, 104.7 ÷ 110.8 = 94.5, and for 1952, 106.3 ÷ 110.8 = 95.9. This is repeated for each index number in the series. The result is shown in Table 10.11.

TABLE 10.11 ILLUSTRATION OF AN INDEX NUMBER SERIES ON AN OLD AND A NEW BASE

Year	Old base: 1950 = 100.0	New base: 1955 = 100.0
1950	100.0	90.3
1951	104.7	94.5
1952	106.3	95.9
1953	109.1	98.5
1954	107.5	97.0
1955	110.8	100.0
1956	113.6	102.5
1957	116.5	105.1
1958	119.2	107.6
1959	120.4	108.7
1960	124.9	112.7
1961	127.6	115.2
1962	128.3	115.8

10.6 DECISIONS TO BE MADE IN CONSTRUCTING AN INDEX NUMBER SERIES

Several basic decisions must be made in planning the construction of an index number series. Among these are the following:

Selection of series to be included: If the series is to be a price index, a careful selection must be made among the variety of goods and price lines available on the market. This involves problems of sampling. In addition, provision must be made for carefully defining the specifications of each commodity included so that data will be comparable from one period to the next.

Selection of data sources: A decision must be made as to whether data are to be collected in the field for each period, which may be daily, weekly, monthly, or quarterly, or whether published sources are to be used. Field collection of data is more costly than use of published data.

Selection of base period: The base period of an index number series should be as close to the present as possible so that it may be possible to compare current data with a period that is relatively close in time. Moreover, the base period selected should be as *normal* as possible. If it is a price index series, then the base period selected should be one that is related to trend, not a period of high or low prices.

Selection of formula and weighting system: Data may be combined by a variety of formulas, using a number of weighting patterns. The pattern selected is dependent upon the objectives of the index number series and how the data are to be used. A good weighting system is basic to the usefulness of an index number series.

10.7 THE CONSUMER PRICE INDEX (TABLE 10.12 AND FIGURE 10.1)[1]

Definition

The Consumer Price Index,[2] known as the CPI, is a statistical measure of changes in prices of goods and services bought by urban wage earners and clerical workers, including both families and single persons living alone. It is not a cost-of-living index, and therefore cannot be used to indicate how much families must actually spend to maintain a specified level of living.

Coverage

The index covers prices of everything people buy for living. It includes food, clothing, shelter, automobiles, housefurnishings, household supplies, fuel, drugs, recreational goods, doctors' and lawyers' fees, beauty shop costs, transportation fares, public utility rates, magazines, and newspapers. It also includes real estate taxes on owned homes, but excludes income and personal property taxes.

[1] The number of published index number series is so large that to review all of them would be a major project. Therefore, two series have been selected for illustrative purposes, the Consumer Price Index and the Wholesale Price Index, both of which are published by the U.S. Department of Labor's Bureau of Labor Statistics.

[2] Based upon U.S. Department of Labor, Bureau of Labor Statistics, *An Abbreviated Description of the Revised Consumer Price Index*, March 3, 1964.

TABLE 10.12 CONSUMER PRICE INDEX, U.S. AVERAGE, ALL ITEMS, 1913–1965

(1957–1959 = 100.0)[a]

Month	1913	1914	1915	1916	1917	1918	1919	1920	1921
Jan	34.2	35.0	35.2	36.4	40.7	48.6	57.6	67.2	66.3
Feb	34.1	34.6	35.0	36.4	41.6	49.1	56.3	68.0	64.2
Mar	34.1	34.5	34.7	36.7	41.9	48.9	57.0	68.7	63.7
Apr	34.2	34.2	35.0	37.1	43.7	49.4	58.0	70.7	62.9
May	34.0	34.4	35.1	37.2	44.7	50.4	58.8	71.9	61.7
Jun	34.1	34.6	35.2	37.7	45.1	51.3	59.0	72.9	61.4
Jul	34.4	35.0	35.2	37.7	44.7	52.6	60.6	72.5	61.5
Aug	34.6	35.5	35.2	38.1	45.4	53.5	61.6	70.6	61.7
Sep	34.8	35.6	35.4	38.8	46.2	54.8	62.0	69.8	61.0
Oct	35.0	35.4	35.7	39.3	47.0	55.7	63.0	69.4	60.9
Nov	35.1	35.5	35.9	40.0	47.1	56.6	64.5	69.0	60.6
Dec	35.0	35.4	36.1	40.2	47.7	57.5	65.9	67.4	60.2
AV	34.5	35.0	35.4	38.0	44.7	52.4	60.3	69.8	62.3

Month	1922	1923	1924	1925	1926	1927	1928	1929	1930
Jan	59.0	58.5	60.1	60.3	62.5	61.0	60.1	59.5	59.6
Feb	58.8	58.4	59.9	60.0	62.2	60.6	59.6	59.4	59.3
Mar	58.2	58.6	59.5	60.1	61.9	60.2	59.6	59.2	59.0
Apr	58.1	58.9	59.3	60.0	62.4	60.2	59.7	59.0	59.3
May	58.1	59.0	59.3	60.2	62.1	60.7	60.0	59.3	59.0
Jun	58.3	59.3	59.3	60.9	61.6	61.3	59.5	59.5	58.6
Jul	58.4	59.9	59.4	61.7	61.0	60.1	59.5	60.1	57.8
Aug	57.8	59.7	59.3	61.7	60.7	59.8	59.7	60.4	57.5
Sep	57.9	60.0	59.6	61.5	61.1	60.1	60.1	60.3	57.8
Oct	58.3	60.1	59.8	61.8	61.3	60.5	60.0	60.3	57.5
Nov	58.5	60.3	60.0	62.8	61.5	60.4	59.9	60.1	57.0
Dec	58.7	60.2	60.1	62.5	61.5	60.3	59.7	59.8	56.2
AV	58.4	59.4	59.6	61.1	61.6	60.5	59.7	59.7	58.2

Month	1931	1932	1933	1934	1935	1936	1937	1938	1939
Jan	55.4	49.8	44.9	46.0	47.4	48.2	49.1	49.6	48.6
Feb	54.6	49.1	44.2	46.4	47.8	47.9	49.2	49.1	48.4
Mar	54.2	48.8	43.8	46.4	47.7	47.7	49.6	49.1	48.3
Apr	53.8	48.5	43.7	46.3	48.1	47.7	49.8	49.3	48.2
May	53.3	47.8	43.9	46.4	47.9	47.7	50.0	49.1	48.2
Jun	52.7	47.5	44.3	46.5	47.8	48.2	50.1	49.1	48.1
Jul	52.6	47.4	45.6	46.5	47.6	48.4	50.3	49.2	48.2
Aug	52.5	46.9	46.0	46.6	47.6	48.7	50.5	49.1	48.1
Sep	52.2	46.6	46.0	47.3	47.8	48.9	50.9	49.1	49.1
Oct	51.9	46.3	46.0	47.0	47.8	48.7	50.7	48.9	48.9
Nov	51.3	46.0	46.0	46.9	48.0	48.7	50.4	48.7	48.8
Dec	50.8	45.6	45.8	46.8	48.2	48.7	50.2	48.8	48.6
AV	53.0	47.6	45.1	46.6	47.8	48.3	50.0	49.1	48.4

(Table 10.12 continued)

TABLE 10.12 (CONTINUED)

(1957–1959 = 100.0)[a]

Month	1940	1941	1942	1943	1944	1945	1946	1947	1948
Jan	48.5	49.1	54.6	58.8	60.6	62.0	63.4	74.9	82.6
Feb	48.8	49.1	55.0	59.0	60.5	61.9	63.2	74.8	81.9
Mar	48.7	49.3	55.7	59.9	60.5	61.9	63.6	76.4	81.7
Apr	48.7	49.8	56.1	60.6	60.8	62.0	64.0	76.4	82.8
May	48.8	50.1	56.6	61.0	61.0	62.5	64.3	76.2	83.4
Jun	49.0	51.0	56.7	60.9	61.2	63.0	65.0	76.8	84.0
Jul	48.9	51.3	57.0	60.5	61.5	63.2	68.9	77.4	85.0
Aug	48.7	51.8	57.3	60.2	61.7	63.2	70.4	78.3	85.4
Sep	48.9	52.7	57.5	60.5	61.8	62.9	71.2	80.1	85.4
Oct	48.8	53.3	58.0	60.7	61.8	62.9	72.6	80.1	85.0
Nov	48.8	53.7	58.4	60.6	61.8	63.2	74.3	80.6	84.4
Dec	49.1	53.9	58.8	60.7	62.0	63.4	74.9	81.7	83.9
AV	48.8	51.3	56.8	60.3	61.3	62.7	68.0	77.8	83.8
Month	1949	1950	1951	1952	1953	1954	1955	1956	1957
Jan	83.7	82.0	88.5	92.2	92.8	93.9	93.2	93.4	96.3
Feb	82.8	81.8	89.6	91.6	92.4	93.7	93.2	93.4	96.7
Mar	83.0	82.1	89.9	91.6	92.6	93.6	93.2	93.5	96.9
Apr	83.2	82.2	90.0	92.0	92.7	93.4	93.1	93.6	97.2
May	83.0	82.6	90.4	92.1	92.9	93.7	93.1	94.1	97.5
Jun	83.1	83.0	90.3	92.4	93.3	93.8	93.2	94.7	98.0
Jul	82.6	83.9	90.4	93.0	93.5	93.9	93.5	95.4	98.5
Aug	82.8	84.5	90.4	93.2	93.7	93.7	93.3	95.2	98.6
Sep	83.2	85.1	91.0	93.0	93.9	93.5	93.6	95.4	98.7
Oct	82.7	85.6	91.4	93.1	94.1	93.3	93.6	95.9	98.7
Nov	82.8	86.0	91.9	93.2	93.7	93.4	93.7	96.0	99.1
Dec	82.3	87.1	92.2	93.0	93.6	93.2	93.5	96.2	99.1
AV	83.0	83.8	90.5	92.5	93.2	93.6	93.3	94.7	98.0
Month	1958	1959	1960	1961	1962	1963	1964	1965	1966
Jan	99.7	100.9	102.2	103.8	104.5	106.0	107.7	108.9	
Feb	99.8	100.8	102.4	103.9	104.8	106.1	107.6	108.9	
Mar	100.5	100.8	102.4	103.9	105.0	106.2	107.7	109.0	
Apr	100.7	101.0	102.9	103.9	105.2	106.2	107.8	109.3	
May	100.7	101.1	102.9	103.8	105.2	106.2	107.8	109.6	
Jun	100.8	101.5	103.1	104.0	105.3	106.6	108.0	110.1	
Jul	101.0	101.8	103.2	104.4	105.5	107.1	108.3	110.2	
Aug	100.8	101.7	103.2	104.3	105.5	107.1	108.2	110.0	
Sep	100.8	102.0	103.3	104.6	106.1	107.1	108.4	110.2	
Oct	100.8	102.3	103.7	104.6	106.0	107.2	108.5	110.4	
Nov	101.0	102.4	103.8	104.6	106.0	107.4	108.7	110.6	
Dec	100.8	102.3	103.9	104.5	105.8	107.6	108.8		
AV	100.7	101.5	103.1	104.2	105.4	106.7	108.1		

[a] As of January 1963, the CPI, formerly calculated on the reference base 1947–49 = 100.0, was converted to the new base, 1957–59 = 100.0 in compliance with recommendations of the U.S. Bureau of the Budget, Office of Statistical Standards.

SOURCE: U.S. Department of Labor, Bureau of Labor Statistics, Washington, D.C.

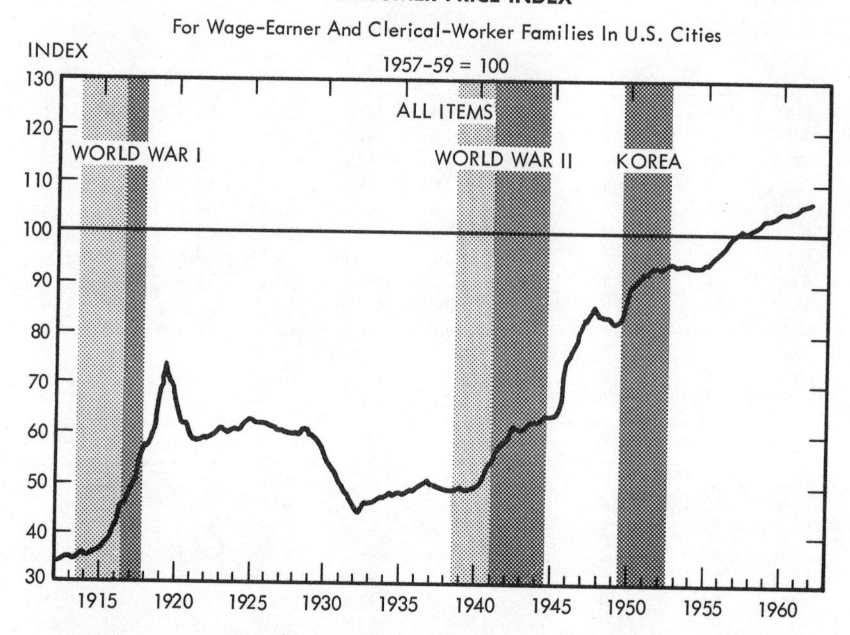

Fig. 10.1 Consumer Price Index, U.S. Average, All Items, 1913–1965
SOURCE: Table 10.11.

Sources of Data

To insure that the index reflects only changes in prices and not changes due to quantity or quality differences, the Bureau of Labor Statistics has prepared detailed specifications to describe the items of the "market basket." Prices are obtained by personal visit from a representative sample of about 16,500 chain stores, independent grocery stores, department and specialty stores, restaurants, professional people, and service establishments where wage and clerical workers buy goods and services. Rental rates are obtained from about 34,000 tenants. Reporters are located both in the city proper and in suburbs of each urban area. Cooperation of reporters is completely voluntary and generally excellent.

The Market Basket

A sample of about 400 items have been selected objectively to compose the "market basket" for current pricing. The list includes all of the most important items bought by consumers, and a sample of the less important ones. The content of this market basket in terms of items, quantities, and qualities is kept essentially unchanged in the index calculation so that any movement of the index from one month to the next is due solely to changes in prices. A comparison of the total cost of the market basket from period to period yields the measure of average price change.

Uses of the Index

The Consumer Price Index is used widely by the general public to guide family budgeting and as an aid in understanding what is happening to family finances. It also is used widely in labor-management contracts to adjust wages. Automatic adjustments based on changes in the index are incorporated in some wage contracts and in a variety of other types of contracts, such as long-term leases. The Consumer Price Index is used as a measure of changes in the purchasing power of the dollar for such purposes as adjusting royalties, pensions, welfare payments, and occasionally alimony payments. It is also used widely as a reflection of inflationary or deflationary trends in the economy.

Limitations

The Consumer Price Index is not an exact measure of price changes. It is subject to sampling errors which cause it to deviate somewhat from the results which would be obtained if actual records of all retail purchases by wage earners and clerical workers could be used to compile the index. These estimating or sampling errors are not mistakes in the index calculation. They are unavoidable. They could be reduced by using much larger samples, but the cost is prohibitive. Furthermore, the index is believed to be sufficiently accurate for most of the practical uses made of it.

Another kind of error occurs because people who give information do not always report accurately. The Bureau of Labor Statistics makes every effort to keep these errors to a minimum, and corrects them whenever they are discovered subsequently. Precautions are taken to guard against errors in pricing, which would affect the index most seriously. The field representatives who collect the price data and the commodity specialists and clerks who process them are well trained to watch for unusual deviations in prices which might be due to errors in reporting.

The Consumer Price Index represents the average movement of prices for wage earners and clerical workers as a broad group, but not necessarily the change in prices paid by any one family or small group of families. The index is not directly applicable to any other occupational group. Some families may find their outlays changing because of changes in factors other than prices, such as family composition. The index measures only the change in prices, and none of the other factors which affect family living expenses.

In many instances, changes in quoted prices are accompanied by changes in the quality of consumer goods and services. Also, new products are introduced frequently which bear little resemblance to products previously on the market. Hence, direct price comparisons cannot be made. The Bureau of Labor Statistics makes every effort to adjust quoted prices for changes in quality, and has developed special procedures for this purpose, including the use of technical specifications and highly trained personnel referred to previously. Nevertheless, some residual effects of quality changes on quoted prices undoubtedly do affect the movement of the Consumer Price Index either downward or upward from time to time.

10.8 THE WHOLESALE PRICE INDEX[1]

Definition

The Wholesale Price Index, known as the WPI, is a statistical measure of price movements in primary markets.

A primary market is one in which the first commercial transaction is made by buyers and sellers, and exists at various stages of production as a commodity is converted from raw material to finished product. Thus, for example, cotton is priced as raw cotton, as yarn, as basic fabrics (gray goods and print cloth), and as finished products (sheets, pillowcases, towels, and shirts). Prices obtained at each of these stages of production are included in the index.

The word "wholesale" as used in the title of the index refers to sales in large lots. It does not refer to prices paid or received by wholesalers, distributors, or jobbers. Most of the price data used in constructing the index are the selling prices of representative manufacturers. The index excludes all retail prices, prices for securities, real estate, services, construction, and transportation. Moreover, the index measures only the effect of price change and not changes in quantity or quality.

Coverage

The Bureau of Labor Statistics compiles three wholesale price indexes: a monthly index, also called comprehensive, Wholesale Price Index, because it includes about 2,000 commodities, and covers the total of transactions at the production level of distribution; a weekly Wholesale Price Index; and a daily index. The range of products included in the Wholesale Price Index is extremely wide, including raw materials, semiprocessed goods,

[1] Based upon Lawrence J. Kaplan, "A Guide to the Federal Government's Indexes of Wholesale Prices," *Analysts Journal*, February 1957.

and manufactured goods. These commodities reflect, directly or indirectly, the sales of all products in the primary markets. Exports are included up to the point at which they leave the domestic market.

Weekly Wholesale Price Index

The weekly wholesale price index represents the best estimate of what the monthly comprehensive index would be if it were computed each week. The weekly index is based on actual prices for fewer than 200 commodities, as compared with approximately 2,000 commodities in the monthly index.

Daily Index of Spot Market Prices

The daily index of spot market prices is designed to measure price movements of 22 selected commodities. These commodities in standardized qualities are traded in a fairly large volume on organized exchanges. This index is a sensitive measure of traders' estimates of current and future economic trends.

Sources of Data

Sources of the price quotations include manufacturers, commodity exchanges, trade journals, and various government agencies. Price schedules are mailed each month to about 5,000 price reporters representing all major industries in all parts of the country. These price data are furnished by businessmen on a purely voluntary basis. Manufacturers provide about 70 per cent of the price quotations, commodity exchanges and trade journals about 20 per cent, and the remainder are obtained from federal government agencies.

Uses of the Monthly Index

The monthly index is used as a barometer of business conditions and as an indicator of the health of the economy. It is also a useful tool in contract escalation when it is necessary to adjust for changes in the prices of the raw and semifinished materials which over a period of time go into the construction of a finished product, such as a ship. The monthly wholesale price index is used as a means of stabilizing the purchasing power of the business dollar in a long-term lease for business, industrial, or commercial property, or a royalty or patent licensing agreement. Purchasing agents and sales managers use the index in comparing the prices they pay for goods, and the movement of their own prices against the index.

Businessmen also use the wholesale price index in budget making and review, in planning the cost of plant expansion programs, in determining depreciation allowances and asset valuation, in market research, in deflation of dollar value series, and in international price comparisons.

Limitations

The index is not a perfect over-all measure of the general price level or the purchasing power of the dollar, even though it is often used to fulfill this function. Moreover, the index measures price movements in primary markets only, and excludes retail transactions as well as jobbing and speculative trading. In spite of these limitations, the wholesale price index reliably measures primary market prices, and thus satisfies the need for which it was constructed.

10.9 THE RELATIONSHIP BETWEEN THE WHOLESALE PRICE INDEX AND THE CONSUMER PRICE INDEX[1]

A comparison of movements in the Bureau of Labor Statistics wholesale (WPI) and consumer price (CPI) indexes over a long period of time and over many selected short periods will show considerable similarity. The reasons for this similarity of movement are twofold. The first is that long-run changes in the level of the WPI involve a somewhat similar change in prices paid by retailers. A second reason is that common cost factors enter into prices for a commodity used by a manufacturer in further production and a commodity purchased by a consumer. For example, the price movement of a textile machine may resemble that of an automobile, because the manufacturers of both pay the same prices for steel and labor.

While the WPI and the CPI show similar movements in the long run, in periods of rapid price changes the wholesale and consumer price indexes diverge considerably, and may even move in opposite directions. The reasons that short-run variations in the index movements may be expected, include, first, the differences in composition of the two indexes. A second reason is that retail-market price changes for the finished goods included in the CPI lag behind primary market price changes for the raw, semi-manufactured, and manufactured goods included in the WPI. And thirdly, short-run variations occur because of the technical difficulties of incorporating into the indexes the prices of goods which are sold at certain seasons.

Differences in Composition

The WPI and CPI differ in what they measure. The CPI is designed to measure the average changes in the retail prices of a fixed, specific market

[1] *Ibid.*

basket or shopping list of goods and services bought by wage-earner and moderate-income households in urban areas. It is not intended to measure all retail prices. The WPI, by contrast, is a measure of price movements at the primary-market level.

Moreover, the WPI is an index of the prices in primary markets of commodities at various stages in the productive process and includes commodities only. Many of these commodities never enter consumer markets. The CPI includes the prices of all the different kinds of things for which consumers spend money, including rents, medical and dental care, and utilities, such as telephone service.

Differences in Time Lag

No consistent time lag exists between the change in price of a raw material and that of a finished product in the primary markets. Similarly, no consistent time lag exists between the change in price of the finished product in the primary market and the change in its retail selling price to the individual consumer. These lags may vary from a few hours to many months.

At one extreme, the price of tomatoes at wholesale at 4:00 A.M. in New York City's wholesale fruit market is reflected at retail all over New York City when stores open at 9:00 A.M. the same morning. If commodities such as this had enough weight in each index, the CPI and WPI would exhibit closely similar movements. At the other extreme, price increases in raw wool may be reflected in the prices of wool apparel items at the clothing manufacturers' level about a year after the rise in price of the raw material. In the case of some meats, price changes at the retail level precede price changes in primary markets.

Seasonal Commodities

Certain commodities, which are sold for seasonal use, raise a technical problem in the indexes. For example, a fur-trimmed coat or an overcoat appears in the stores late in the summer or early in the fall. The Bureau of Labor Statistics must incorporate in its indexes the price changes between the end of one season and the beginning of the next. Although no real market price exists during the off season, price changes are accruing even when the item is off the market. End-of-season prices are generally lowered in order to dispose of seasonal merchandise, and early-season prices tend to be as high as the traffic will bear.

The price problem of seasonal commodities, however, is largely a matter of time differences. The practice of the Bureau of Labor Statistics is to incorporate these accruing prices at the beginning of the new season—a time that is later for the CPI than for the WPI. In the month when such a change enters into the CPI, a divergence may be created from the WPI, and vice versa.

EXERCISES

1. Define or explain the following:

Aggregative index	Laspeyres formula
Arithmetic average of price relatives	Market basket
	Marshall-Edgeworth formula
Base year	Paasche formula
Base-year weights	Price indexes
Changing the base year	Price relative
Consumer Price Index	Simple index
Farm price indexes	Special-purpose index numbers
Fisher's ideal index	Splicing
Given-year weights	Value indexes
Index number	Value weights
Index of aggregate prices	Weighted index
Index of physical quantity	Wholesale Price Index

2. Given the data in Table 10.13, and using 1956 = 100.0 as the base year:
   a. Construct a simple arithmetic average of price relatives.
   b. Construct a simple aggregate of actual prices.
   c. Construct a weighted aggregative index, weighted by base-year quantities.

TABLE 10.13 ANNUAL AVERAGE PRICES FOR COMMODITIES A, B, AND C, 1956, 1960, 1962, AND QUANTITIES PRODUCED, 1956

Commodity	Prices			Quantity produced 1956 (Millions of units)
	1956	1960	1962	
A	$1.00	$1.54	$1.62	225
B	0.50	0.76	0.83	130
C	0.80	1.21	1.32	178
Total	$2.30	$3.51	$3.77	xxx

3. Given the two index number series, representing production of the XYZ Manufacturing Company, in Table 10.14, splice the two series.

TABLE 10.14 PRODUCTION OF THE XYZ MANUFACTURING COMPANY, 1954–1962

Year	Production	
	1954 = 100.0 base	1958 = 100.0 base
1954	100.0	
1955	106.8	
1956	108.4	
1957	111.2	
1958	109.6	100.0
1959		103.6
1960		105.3
1961		106.2
1962		108.0

4. Given the data in Table 10.15,

TABLE 10.15 INDEX OF SALES VOLUME, AJAX MANUFACTURING COMPANY, 1956–1964
(1956 = 100.0)

Year	Sales volume
1956	100.0
1957	107.9
1958	109.5
1959	112.4
1960	110.8
1961	114.2
1962	116.9
1963	119.8
1964	123.7

change the base from 1956 = 100.0 to 1960 = 100.0.

5. List the uses and limitations of the Consumer Price Index and the Wholesale Price Index.

6. What is the relationship between the Wholesale Price Index and the Consumer Price Index?

APPENDIX **A**

How To Use Logarithms

Routine calculations can be made by the use of simple arithmetic, a desk calculator, the slide rule, or logarithms. The purpose of this Appendix is to review the use of logarithms.

Logarithms are a tool useful for making routine calculations, such as multiplication, division, and extraction of roots. Moreover, logarithms are important to the statistician for other reasons.

1. Logarithms are the basis for the slide rule.
2. Logarithms provide the basis for logarithmic graph paper.
3. Logarithms are a component of a few statistical formulas.

SYSTEMS OF LOGARITHMS

Tables of logarithms of numbers can be made up for any base. When logarithms were first given to the world by Lord Napier in 1614, the base of his system involved the constant *e* which has the value 2.71828. These are called Napierian logarithms and, sometimes, natural logarithms.

Briggs, another mathematician, introduced 10 as a base for logarithms. He reasoned that since our number system has the base 10, multiplication and division by the use of powers of 10 would be easier and more convenient than the use of any other value. The following discussion of logarithms uses 10 as base.

DEFINITION

The *logarithm of a number* to the base 10 is the power to which 10 must be raised to produce the number.

$$\text{The log of } 1{,}000 \text{ to base } 10 = 3$$
$$\text{or } (10)\,(10)\,(10) = 10^3 = 1{,}000$$

The logarithm is the exponent 3.

Examples

The logs of some common numbers to the base 10 are as follows:

$$\begin{aligned}
\log 1 &= 0\\
\log 2 &= 0.3010\\
\log 20 &= 1.3010\\
\log 200 &= 2.3010\\
\log 500 &= 2.6990\\
\log 515 &= 2.7118\\
\log 1{,}000 &= 3.0000\\
\log 0.2 &= 9.3010 - 10\\
\log 0.02 &= 8.3010 - 10\\
\log 0.002 &= 7.3010 - 10
\end{aligned}$$

CHARACTERISTIC AND MANTISSA

Each logarithm is made up of two parts, an integer and a decimal fraction. The integer is called the *characteristic* of the logarithm, and the decimal fraction is called the *mantissa.*

Characteristic

When a number is larger than 1 its characteristic is one less than the number of digits to the left of the decimal point in the number.

When a number is smaller than 1 its characteristic is negative, and is numerically one more than the number of zeros between the decimal point and the first significant digit in the number.

Mantissa

The mantissa is obtained from a table of common logarithms, for example, Table I, on pages 251 and 252. The mantissas are in the body of the table, and the numbers are in the margins. The number corresponding to a given logarithm is called its *antilogarithm* or *antilog.*

TABLE I COMMON LOGARITHMS

N	0	1	2	3	4	5	6	7	8	9
10	0000	0043	0086	0128	0170	0212	0253	0294	0334	0374
11	0414	0453	0492	0531	0569	0607	0645	0682	0719	0755
12	0792	0828	0864	0899	0934	0969	1004	1038	1072	1106
13	1139	1173	1206	1239	1271	1303	1335	1367	1399	1430
14	1461	1492	1523	1553	1584	1614	1644	1673	1703	1732
15	1761	1790	1818	1847	1875	1903	1931	1959	1987	2014
16	2041	2068	2095	2122	2148	2175	2201	2227	2253	2279
17	2304	2330	2355	2380	2405	2430	2455	2480	2504	2529
18	2553	2577	2601	2625	2648	2672	2695	2718	2742	2765
19	2788	2810	2833	2856	2878	2900	2923	2945	2967	2989
20	3010	3032	3054	3075	3096	3118	3139	3160	3181	3201
21	3222	3243	3263	3284	3304	3324	3345	3365	3385	3404
22	3424	3444	3464	3483	3502	3522	3541	3560	3579	3598
23	3617	3636	3655	3674	3692	3711	3729	3747	3766	3784
24	3802	3820	3838	3856	3874	3982	3909	3927	3945	3962
25	3979	3997	4014	4031	4048	4065	4082	4099	4116	4133
26	4150	4166	4183	4200	4216	4232	4249	4265	4281	4298
27	4314	4330	4346	4362	4378	4393	4409	4425	4440	4456
28	4472	4487	4502	4518	4533	4548	4564	4579	4594	4609
29	4624	4639	4654	4669	4683	4698	4713	4728	4742	4757
30	4771	4786	4800	4814	4829	4843	4857	4871	4886	4900
31	4914	4928	4942	4955	4969	4983	4997	5011	5024	5038
32	5051	5065	5079	5092	5105	5119	5132	5145	5159	5172
33	5185	5198	5211	5224	5237	5250	5263	5276	5289	5302
34	5315	5328	5340	5353	5366	5378	5391	5403	5416	5428
35	5441	5453	5465	5478	5490	5502	5514	5527	5539	5551
36	5563	5575	5587	5599	5611	5623	5635	5647	5658	5670
37	5682	5694	5705	5717	5729	5740	5752	5763	5775	5786
38	5798	5809	5821	5832	5843	5855	5866	5877	5888	5899
39	5911	5922	5933	5944	5955	5966	5977	5988	5999	6010
40	6021	6031	6042	6053	6064	6075	6085	6096	6107	6117
41	6128	6138	6149	6160	6170	6180	6191	6201	6212	6222
42	6232	6243	6253	6263	6274	6284	6294	6304	6314	6325
43	6335	6345	6355	6365	6375	6385	6395	6405	6415	6425
44	6435	6444	6454	6464	6474	6484	6493	6503	6513	6522
45	6532	6542	6551	6561	6571	6580	6590	6599	6609	6618
46	6628	6637	6646	6656	6665	6675	6684	6693	6702	6712
47	6721	6730	6739	6749	6758	6767	6776	6785	6794	6803
48	6812	6821	6830	6839	6848	6857	6866	6875	6884	6893
49	6902	6911	6920	6928	6937	6946	6955	6964	6972	6981
50	6990	6998	7007	7016	7024	7033	7042	7050	7059	7067
51	7076	7084	7093	7101	7110	7118	7126	7135	7143	7152
52	7160	7168	7177	7185	7193	7202	7210	7218	7226	7235
53	7243	7251	7259	7267	7275	7284	7292	7300	7308	7316
54	7324	7332	7340	7348	7356	7364	7372	7380	7388	7396
N	0	1	2	3	4	5	6	7	8	9

(Table I continued)

TABLE I COMMON LOGARITHMS (CONTINUED)

N	0	1	2	3	4	5	6	7	8	9
55	7404	7412	7419	7427	7435	7443	7451	7459	7466	7474
56	7482	7490	7497	7505	7513	7520	7528	7536	7543	7551
57	7559	7566	7574	7582	7589	7597	7604	7612	7619	7627
58	7634	7642	7649	7657	7664	7672	7679	7686	7694	7701
59	7709	7716	7723	7731	7738	7745	7752	7760	7767	7774
60	7782	7789	7796	7803	7810	7818	7825	7832	7839	7846
61	7853	7860	7868	7875	7882	7889	7896	7903	7910	7917
62	7924	7931	7938	7945	7952	7959	7966	7973	7980	7987
63	7993	8000	8007	8014	8021	8028	8035	8041	8048	8055
64	8062	8069	8075	8082	8089	8096	8102	8109	8116	8122
65	8129	8136	8142	8149	8156	8162	8169	8176	8182	8189
66	8195	8202	8209	8215	8222	8228	8235	8241	8248	8254
67	8261	8267	8274	8280	8287	8293	8299	8306	8312	8319
68	8325	8331	8338	8344	8351	8357	8363	8370	8376	8382
69	8388	8395	8401	8407	8414	8420	8426	8432	8439	8445
70	8451	8457	8463	8470	8476	8482	8488	8494	8500	8506
71	8513	8519	8525	8531	8537	8543	8549	8555	8561	8567
72	8573	8579	8585	8591	8597	8603	8609	8615	8621	8627
73	8633	8639	8645	8651	8657	8663	8669	8675	8681	8686
74	8692	8698	8704	8710	8716	8722	8727	8733	8739	8745
75	8751	8756	8762	8768	8774	8779	8785	8791	8797	8802
76	8808	8814	8820	8825	8831	8837	8842	8848	8854	8859
77	8865	8871	8876	8882	8887	8893	8899	8904	8910	8915
78	8921	8927	8932	8938	8943	8949	8954	8960	8965	8971
79	8976	8982	8987	8993	8998	9004	9009	9015	9020	9025
80	9031	9036	9042	9047	9053	9058	9063	9069	9074	9079
81	9085	9090	9096	9101	9106	9112	9117	9122	9128	9133
82	9138	9143	9149	9154	9159	9165	9170	9175	9180	9186
83	9191	9196	9201	9206	9212	9217	9222	9227	9232	9238
84	9243	9248	9253	9258	9263	9269	9274	9279	9284	9289
85	9294	9299	9304	9309	9315	9320	9325	9330	9335	9340
86	9345	9350	9355	9360	9365	9370	9375	9380	9385	9390
87	9395	9400	9405	9410	9415	9420	9425	9430	9435	9440
88	9445	9450	9455	9460	9465	9469	9474	9479	9484	9489
89	9494	9499	9504	9509	9513	9518	9523	9528	9533	9538
90	9542	9547	9552	9557	9562	9566	9571	9576	9581	9586
91	9590	9595	9600	9605	9609	9614	9619	9624	9628	9633
92	9638	9643	9647	9652	9657	9661	9666	9671	9675	9680
93	9685	9689	9694	9699	9703	9708	9713	9717	9722	9727
94	9731	9736	9741	9745	9750	9754	9759	9763	9768	9773
95	9777	9782	9786	9791	9795	9800	9805	9809	9814	9818
96	9823	9827	9832	9836	9841	9845	9850	9854	9859	9863
97	9868	9872	9877	9881	9886	9890	9894	9899	9903	9908
98	9912	9917	9921	9926	9930	9934	9939	9943	9948	9952
99	9956	9961	9965	9969	9974	9978	9983	9987	9991	9996
N	0	1	2	3	4	5	6	7	8	9

APPLICATION

The characteristic and mantissa for 628:

The characteristic is 2. The number 628 has 3 digits, and one less than the number of digits is 2.

To find the mantissa of the logarithm of 628, locate 62 in the left-hand vertical column headed N and 8 in the horizontal row across the top of the table. The mantissa is in the cell of the table corresponding to these two entries. The mantissa is 7980. Therefore, log 628 = 2.7980, or $10^{2.7980} = 628$.

RULES FOR COMPUTATION

1. To obtain a product, add the logarithms of the factors.

$$A \times B \rightarrow \log A + \log B$$

2. To obtain a quotient, subtract the logarithm of the denominator from the logarithm of the numerator.

$$A \div B \rightarrow \log A - \log B$$

3. To obtain the value of a number raised to a power, multiply the power of the number by the logarithm of the number.

$$A^n \rightarrow n \log A$$

4. To obtain the nth root of a number, divide the logarithm of the number by n.

$$\sqrt[n]{A} \rightarrow \frac{\log A}{n}$$

In each case, the final step requires that the logarithm of the operation indicated above be transformed back to the natural number.

Example of Each Rule

1. $$(575)(342) \rightarrow \log 575 + \log 342$$

2. $$\frac{575}{342} \rightarrow \log 575 - \log 342$$

3. $$575^8 \rightarrow (8)(\log 575)$$

4. $$\sqrt[8]{575} \rightarrow \frac{\log 575}{8}$$

INTERPOLATION

The table of mantissas gives the logarithms of three-place numbers. The logarithm of a four-place number may be computed by interpolation.

EXAMPLE 1: Find the logarithm of 4,362.

(1) Number	(2) Mantissa
4,370	6405
4,362	x
4,360	6395

Column 1: Subtract 4,360 from 4,362 to obtain 2.
Then, subtract 4,360 from 4,370 to obtain 10.
Form a ratio: 2/10.

Column 2: Subtract 6395 from x to obtain x.
Then, subtract 6395 from 6405 to obtain 10.
Form a ratio: $x/10$.

Solve for x:

$$\frac{2}{10} = \frac{x}{10}$$

$$10x = 20$$

$$x = 2$$

Add 2 to 6395 to obtain 3.6397. Thus, log 4,362 = 3.6397.

EXAMPLE 2: Find the antilog of 3.6397.

(1)	(2)
4,370	6405
x	6397
4,360	6395

$$\frac{x}{10} = \frac{2}{10}$$

$$10x = 20$$

$$x = 2$$

Antilog of 3.6397 = 4,362

SAMPLE PROBLEM:

Given:

$$x = \frac{(515)(214)}{0.214}$$

Solve by the use of logarithms.

Solution

$$\log x = \log 515 + \log 214 - \log 0.214$$
$$\log x = 2.7118 + 2.3304 - (9.3304 - 10)$$
$$\log x = 2.7118 + 2.3304 - 9.3304 + 10$$
$$\log x = 5.7118$$
$$x = 515{,}000$$

Note:

$$\begin{array}{r} 2.7118 \\ +\ \underline{2.3304} \\ 5.0422 \end{array}$$

$$\begin{array}{r} 5.0422 \\ \underline{-(+9.3304 - 10)} \end{array}$$

$$\begin{array}{r} 15.0422 \cancel{- 10} \\ \underline{-9.3304 \cancel{+ 10}} \\ 5.7118 \end{array}$$

APPENDIX **B**

Steps in Computation of Square Root

THE CONCEPT

The square root of 16 is + 4, or − 4, since (+4) (+4) = 16, and (−4) (−4) = 16. The *principal square root* of a number is its positive square root, and is the root generally called for in a statistical problem concerned with an amount, a quantity, or a magnitude.

The square root of 16 is written as $\sqrt[2]{16}$ or simply $\sqrt{16}$. In the expression $\sqrt{16}$, 16 is called the *radicand*, which is the number for which the square root is to be found. $\sqrt{\ }$ is called the *radical sign*. Thus, $\sqrt{16}$ means, "What number multiplied by itself equals 16?" Hence, $\sqrt{16} = +4$ or -4.

DEFINITION: The square root of a number is one of its two equal factors. Every number has two square roots.

EXAMPLE: Find the square root of 3749.8625.

$$
\begin{array}{lr}
 & 6\ \ 1.\ \ 2\ \ 3 \\
 & \sqrt{\overset{\frown}{37}\overset{\frown}{49}.\overset{\frown}{86}\overset{\frown}{25}} \\
 & 36 \\
121 & 149 \\
 & 121 \\
1222 & 28\ 86 \\
 & 24\ 44 \\
12243 & 4\ 4225 \\
 & 3\ 6729
\end{array}
$$

STEPS

1. Place decimal point above the radical sign. Then, starting at the decimal point in the radicand, pair off the digits in both directions as indicated by the arcs.

2. The largest number whose square is less than the first pair, 37, is 6. Hence 6 is placed above the 37. Next, subtract its square, 36, from the 37. Then bring down the next pair, 49, as shown.

3. Double the 6, obtaining the 12 shown. Block out momentarily the last digit, 9, in 149; 12 will go into 14 at most 1 time. Hence, write the 1 above the second pair and adjacent to the 12 as shown.

4. Multiply 121 by 1 and subtract the result from 149, obtaining 28. Bring down the next pair of digits, obtaining 2886.

5. Doubling 61 gives 122. Block out the last digit in 2886; 122 will go into 288 at most 2 times. Write a 2 above the third group and beside the 122 as shown.

6. Multiply 1222 by 2. Subtract 2444 from 2886, leaving a remainder of 442. Bring down the next pair to obtain 44225.

7. Doubling 612 gives 1224, which goes into 4422 at most 3 times. Write 3 above and adjacent to 1224, as shown.

8. Multiply 12243 by 3, and subtract the result from 44225.

9. This procedure may be continued by adding pairs of zeros after the last pair in the radicand to obtain any number of significant figures, depending upon the degree of accuracy required.

10. In this example, the answer to four significant figures is $\sqrt{3749.8625} = 61.23$.

USING A TABLE TO OBTAIN SQUARES AND SQUARE ROOTS OF NUMBERS

Table II provides squares and square roots for numbers from 1 to 1,000. The table is used as shown on pp. 259–271.

EXAMPLE 1

Obtain the principal square root of 78 to the nearest tenth by using the table.

Solution: Locate 78 in the number column of the table. In the square root column, the number is 8.83. The principal square root of 78 correct to the nearest tenth is 8.8.

EXAMPLE 2

Obtain the square of 135 by using the table.

Solution: Locate 135 in the number column of the table. In the square column, the number is 18,225. The square of 135 is 18,225.

TABLE II SQUARES AND SQUARE ROOTS

Number	Square	Square root	Number	Square	Square root
1	1	1.0000	41	1,681	6.4031
2	4	1.4142	42	1,764	6.4807
3	9	1.7321	43	1,849	6.5574
4	16	2.0000	44	1,936	6.6332
5	25	2.2361	45	2,025	6.7082
6	36	2.4495	46	2,116	6.7823
7	49	2.6458	47	2,209	6.8557
8	64	2.8284	48	2,304	6.9282
9	81	3.0000	49	2,401	7.0000
10	100	3.1623	50	2,500	7.0711
11	121	3.3166	51	2,601	7.1414
12	144	3.4641	52	2,704	7.2111
13	169	3.6056	53	2,809	7.2801
14	196	3.7417	54	2,916	7.3485
15	225	3.8730	55	3,025	7.4162
16	256	4.0000	56	3,136	7.4833
17	289	4.1231	57	3,249	7.5498
18	324	4.2426	58	3,364	7.6158
19	361	4.3589	59	3,481	7.6811
20	400	4.4721	60	3,600	7.7460
21	441	4.5826	61	3,721	7.8102
22	484	4.6904	62	3,844	7.8740
23	529	4.7958	63	3,969	7.9373
24	576	4.8990	64	4,096	8.0000
25	625	5.0000	65	4,225	8.0623
26	676	5.0990	66	4,356	8.1240
27	729	5.1962	67	4,489	8.1854
28	784	5.2915	68	4,624	8.2462
29	841	5.3852	69	4,761	8.3066
30	900	5.4772	70	4,900	8.3666
31	961	5.5678	71	5,041	8.4261
32	1,024	5.6569	72	5,184	8.4853
33	1,089	5.7446	73	5,329	8.5440
34	1,156	5.8310	74	5,476	8.6023
35	1,225	5.9161	75	5,625	8.6603
36	1,296	6.0000	76	5,776	8.7178
37	1,369	6.0828	77	5,929	8.7750
38	1,444	6.1644	78	6,084	8.8318
39	1,521	6.2450	79	6,241	8.8882
40	1,600	6.3246	80	6,400	8.9443

TABLE II SQUARES AND SQUARE ROOTS (CONTINUED)

Number	Square	Square root	Number	Square	Square root
81	6,561	9.0000	121	14,641	11.0000
82	6,724	9.0554	122	14,884	11.0454
83	6,889	9.1104	123	15,129	11.0905
84	7,056	9.1652	124	15,376	11.1355
85	7,225	9.2195	125	15,625	11.1803
86	7,396	9.2736	126	15,876	11.2250
87	7,569	9.3274	127	16,129	11.2694
88	7,744	9.3808	128	16,384	11.3137
89	7,921	9.4340	129	16,641	11.3578
90	8,100	9.4868	130	16,900	11.4018
91	8,281	9.5394	131	17,161	11.4455
92	8,464	9.5917	132	17,424	11.4891
93	8,649	9.6437	133	17,689	11.5326
94	8,836	9.6954	134	17,956	11.5758
95	9.025	9.7468	135	18,225	11.6190
96	9,216	9.7980	136	18,496	11.6619
97	9,409	9.8489	137	18,769	11.7047
98	9,604	9.8995	138	19,044	11.7473
99	9,801	9.9499	139	19,321	11.7898
100	10,000	10.000	140	19,600	11.8322
101	10,201	10.0499	141	19,881	11.8743
102	10,404	10.0995	142	20,164	11.9164
103	10,609	10.1489	143	20,449	11.9583
104	10,816	10.1980	144	20,736	12.0000
105	11,025	10.2470	145	21,025	12.0416
106	11,236	10.2956	146	21,316	12.0830
107	11,449	10.3441	147	21.609	12.1244
108	11,664	10.3923	148	21,904	12.1655
109	11,881	10.4403	149	22,201	12.2066
110	12,100	10.4881	150	22,500	12.2474
111	12,321	10.5357	151	22,801	12.2882
112	12,544	10.5830	152	23,104	12.3288
113	12,769	10.6301	153	23,409	12.3693
114	12,996	10.6771	154	23,716	12.4097
115	13,225	10.7238	155	24,025	12.4499
116	13,456	10.7703	156	24,336	12.4900
117	13,689	10.8167	157	24,649	12.5300
118	13,924	10.8628	158	24,964	12.5698
119	14,161	10.9087	159	25,281	12.6095
120	14,400	10.9545	160	25,600	12.6491

TABLE II SQUARES AND SQUARE ROOTS (CONTINUED)

Number	Square	Square root	Number	Square	Square root
161	25,921	12.6886	201	40,401	14.1774
162	26,244	12.7279	202	40,804	14.2127
163	26,569	12.7671	203	41,209	14.2478
164	26,896	12.8062	204	41,616	14.2829
165	27,225	12.8452	205	42,025	14.3178
166	27,556	12.8841	206	42,436	14.3527
167	27,889	12.9228	207	42,849	14.3875
168	28,224	12.9615	208	43,264	14.4222
169	28,561	13.0000	209	43,681	14.4568
170	28,900	13.0384	210	44,100	14.4914
171	29,241	13.0767	211	44,521	14.5258
172	29,584	13.1149	212	44,944	14.5602
173	29,929	13.1529	213	45,369	14.5945
174	30,276	13.1909	214	45,796	14.6287
175	30,625	13.2288	215	46,225	14.6629
176	30,976	13.2665	216	46,656	14.6969
177	31,329	13.3041	217	47,089	14.7309
178	31,684	13.3417	218	47,524	14.7648
179	32,041	13.3791	219	47,961	14.7986
180	32,400	13.4164	220	48,400	14.8324
181	32,761	13.4536	221	48,841	14.8661
182	33,124	13.4907	222	49,284	14.8997
183	33,489	13.5277	223	49,729	14.9332
184	33,856	13.5647	224	50,176	14.9666
185	34,225	13.6015	225	50,625	15.0000
186	34,596	13.6382	226	51,076	15.0333
187	34,969	13.6748	227	51,529	15.0665
188	35,344	13.7113	228	51,984	15.0997
189	35,721	13.7477	229	52,441	15.1327
190	36,100	13.7840	230	52,900	15.1658
191	36,481	13.8203	231	53,361	15.1987
192	36,864	13.8564	232	53,824	15.2315
193	37,249	13.8924	233	54,289	15.2643
194	37,636	13.9284	234	54,756	15.2971
195	38,025	13.9642	235	55,225	15.3297
196	38,416	14.0000	236	55,696	15.3623
197	38,809	14.0357	237	56,169	15.3948
198	39,204	14.0712	238	56,644	15.4272
199	39,601	14.1067	239	57,121	15.4596
200	40,000	14.1421	240	57,600	15.4919

TABLE II SQUARES AND SQUARE ROOTS (CONTINUED)

Number	Square	Square root	Number	Square	Square root
241	58,081	15.5242	281	78,961	16.7631
242	58,564	15.5563	282	79,524	16.7929
243	59,049	15.5885	283	80,089	16.8226
244	59,536	15.6205	284	80,656	16.8523
245	60,025	15.6525	285	81,225	16.8819
246	60,516	15.6844	286	81,796	16.9115
247	61,009	15.7162	287	82,369	16.9411
248	61,504	15.7480	288	82,944	16.9706
249	62,001	15.7797	289	83,521	17.0000
250	62,500	15.8114	290	84,100	17.0294
251	63,001	15.8430	291	84,681	17.0587
252	63,504	15.8745	292	85,264	17.0880
253	64,009	15.9060	293	85,849	17.1172
254	64,516	15.9374	294	86,436	17.1464
255	65,025	15.9687	295	87,025	17.1756
256	65,536	16.0000	296	87,616	17.2047
257	66,049	16.0312	297	88,209	17.2337
258	66,564	16.0624	298	88,804	17.2627
259	67,081	16.0935	299	89,401	17.2916
260	67,600	16.1245	300	90,000	17.3205
261	68,121	16.1555	301	90,601	17.3494
262	68,644	16.1864	302	91,204	17.3781
263	69,169	16.2173	303	91,809	17.4069
264	69,696	16.2481	304	92,416	17.4356
265	70,225	16.2788	305	93,025	17.4642
266	70,756	16.3095	306	93,636	17.4929
267	71,289	16.3401	307	94,249	17.5214
268	71,824	16.3707	308	94,864	17.5499
269	72,361	16.4012	309	95,481	17.5784
270	72,900	16.4317	310	96,100	17.6068
271	73,441	16.4621	311	96,721	17.6352
272	73,984	16.4924	312	97,344	17.6635
273	74,529	16.5227	313	97,969	17.6918
274	75,076	16.5529	314	98,596	17.7200
275	75,625	16.5831	315	99,225	17.7482
276	76,176	16.6132	316	99,856	17.7764
277	76,729	16.6433	317	100,489	17.8045
278	77,284	16.6733	318	101,124	17.8326
279	77,841	16.7033	319	101,761	17.8606
280	78,400	16.7332	320	102,400	17.8885

TABLE II SQUARES AND SQUARE ROOTS (CONTINUED)

Number	Square	Square root	Number	Square	Square root
321	103,041	17.9165	361	130,321	19.0000
322	103,684	17.9444	362	131,044	19.0263
323	104,329	17.9722	363	131,769	19.0526
324	104,976	18.0000	364	132,496	19.0788
325	105,625	18.0278	365	133,225	19.1050
326	106,276	18.0555	366	133,956	19.1311
327	106,929	18.0831	367	134,689	19.1572
328	107,584	18.1108	368	135,424	19.1833
329	108,241	18.1384	369	136,161	19.2094
330	108,900	18.1659	370	136,900	19.2354
331	109,561	18.1934	371	137,641	19.2614
332	110,224	18.2209	372	138,384	19,2873
333	110,889	18.2483	373	139,129	19.3132
334	111,556	18.2757	374	139,876	19.3391
335	112,225	18.3030	375	140,625	19.3649
336	112,896	18.3303	376	141,376	19.3907
337	113,569	18.3576	377	142,129	19.4165
338	114,244	18.3848	378	142,884	19.4422
339	114,921	18.4120	379	143,641	19.4679
340	115,600	18.4391	380	144,400	19.4936
341	116,281	18.4662	381	145,161	19.5192
342	116,964	18.4932	382	145,924	19.5448
343	117,649	18.5203	383	146,689	19.5704
344	118,336	18.5472	384	147,456	19.5959
345	119,025	18.5742	385	148,225	19.6214
346	119,716	18.6011	386	148,996	19.6469
347	120,409	18.6279	387	149,769	19.6723
348	121,104	18.6548	388	150,544	19.6977
349	121,801	18.6815	389	151,321	19.7231
350	122,500	18.7083	390	152,100	19.7484
351	123,201	18.7350	391	152,881	19.7737
352	123,904	18.7617	392	153,664	19.7990
353	124,609	18.7883	393	154,449	19.8242
354	125,316	18.8149	394	155,236	19.8494
355	126,025	18.8414	395	156,025	19.8746
356	126,736	18.8680	396	156,816	19.8997
357	127,449	18.8944	397	157,609	19.9249
358	128,164	18.9209	398	158,404	19.9499
359	128,881	18.9473	399	159,201	19.9750
360	129,600	18.9737	400	160,000	20.0000

TABLE II SQUARES AND SQUARE ROOTS (CONTINUED)

Number	Square	Square root	Number	Square	Square root
401	160,801	20.0250	441	194,481	21.0000
402	161,604	20.0499	442	195,364	21.0238
403	162,409	20.0749	443	196,249	21.0476
404	163,216	20.0998	444	197,136	21.0713
405	164,025	20.1246	445	198,025	21.0950
406	164,836	20.1494	446	198,916	21.1187
407	165,649	20.1742	447	199,809	21.1424
408	166,464	20.1990	448	200,704	21.1660
409	167,281	20.2237	449	201,601	21.1896
410	168,100	20.2485	450	202,500	21.2132
411	168,921	20.2731	451	203,401	21.2368
412	169,744	20.2978	452	204,304	21.2603
413	170,569	20.3224	453	205,209	21.2838
414	171,396	20.3470	454	206,116	21.3073
415	172,225	20.3715	455	207,025	21.3307
416	173,056	20.3961	456	207,936	21.3542
417	173,889	20.4206	457	208,849	21.3776
418	174,724	20.4450	458	209,764	21.4009
419	175,561	20.4695	459	210,681	21.4243
420	176,400	20.4939	460	211,600	21.4476
421	177,241	20.5183	461	212,521	21.4709
422	178,084	20.5426	462	213,444	21.4942
423	178,929	20.5670	463	214,369	21.5174
424	179,776	20.5913	464	215,296	21.5407
425	180,625	20.6155	465	216,225	21.5639
426	181,476	20.6398	466	217,156	21.5870
427	182,329	20.6640	467	218,089	21.6102
428	183,184	20.6882	468	219,024	21.6333
429	184,041	20.7123	469	219,961	21.6574
430	184,900	20.7364	470	220,900	21.6795
431	185,761	20.7605	471	221,841	21.7025
432	186,624	20.7846	472	222,784	21.7256
433	187,489	20.8087	473	223,729	21.7486
434	188,356	20.8327	474	224,676	21.7715
435	189,225	20.8567	475	225,625	21.7945
436	190,096	20.8806	476	226,576	21.8174
437	190,969	20.9045	477	227,529	21.8403
438	191,844	20.9284	478	228,484	21.8632
439	192,721	20.9523	479	229,441	21.8861
440	193,600	20.9762	480	230,400	21.9089

TABLE II SQUARES AND SQUARE ROOTS (CONTINUED)

Number	Square	Square root	Number	Square	Square root
481	231,361	21.9317	521	271,441	22.8254
482	232,324	21.9545	522	272,484	22.8473
483	233,289	21.9773	523	273,529	22.8692
484	234,256	22.0000	524	274,576	22.8910
485	235,225	22.0227	525	275,625	22.9129
486	236,196	22.0454	526	276,676	22.9347
487	237,169	22.0681	527	277,729	22.9565
488	238,144	22.0907	528	278,784	22.9783
489	239,121	22.1133	529	279,841	23.0000
490	240,100	22.1359	530	280,900	23.0217
491	241,081	22.1585	531	281,961	23.0434
492	242,064	22.1811	532	283,024	23.0651
493	243,049	22.2036	533	284,089	23.0868
494	244,036	22.2261	534	285,156	23.1084
495	245,025	22.2486	535	286,225	23.1301
496	246,016	22.2711	536	287,296	23.1517
497	247,009	22.2935	537	288,369	23.1733
498	248,004	22.3159	538	289,444	23.1948
499	249,001	22.3383	539	290,521	23.2164
500	250,000	22.3607	540	291,600	23.2379
501	251,001	22.3830	541	292,681	23.2594
502	252,004	22.4054	542	293,764	23.2809
503	253,009	22.4277	543	294,849	23.3024
504	254,016	22.4499	544	295,936	23.3238
505	255,025	22.4722	545	297,025	23.3452
506	256,036	22.4944	546	298,116	23.3666
507	257,049	22.5167	547	299,209	23.3880
508	258,064	22.5389	548	300,304	23.4094
509	259,081	22.5610	549	301,401	23.4307
510	260,100	22.5832	550	302,500	23.4521
511	261,121	22.6053	551	303,601	23.4734
512	262,144	22.6274	552	304,704	23.4947
513	263,169	22.6495	553	305,809	23.5160
514	264,196	22.6716	554	306,916	23.5372
515	265,225	22.6936	555	308,025	23.5584
516	266,256	22.7156	556	309,136	23.5797
517	267,289	22.7376	557	310,249	23.6008
518	268,324	22.7596	558	311,364	23.6220
519	269,361	22.7816	559	312,481	23.6432
520	270,400	22.8035	560	313,600	23.6643

TABLE II SQUARES AND SQUARE ROOTS (CONTINUED)

Number	Square	Square root	Number	Square	Square root
561	314,721	23.6854	601	361,201	24.5153
562	315,844	23.7065	602	362,404	24.5357
563	316,969	23.7276	603	363,609	24.5561
564	318,096	23.7487	604	364,816	24.5764
565	319,225	23.7697	605	366,025	24.5967
566	320,356	23.7908	606	367,236	24.6171
567	321,489	23.8118	607	368,449	24.6374
568	322,624	23.8328	608	369,664	24.6577
569	323,761	23.8537	609	370,881	24.6779
570	324,900	23.8747	610	372,100	24.6982
571	326,041	23.8956	611	373,321	24.7184
572	327,184	23.9165	612	374,544	24.7385
573	328,329	23.9374	613	375,769	24.7588
574	329,476	23.9583	614	376,996	24.7790
575	330,625	23.9792	615	378,225	24.7992
576	331,776	24.0000	616	379,456	24.8193
577	332,929	24.0208	617	380,689	24.8395
578	334,084	24.0416	618	381,924	24.8596
579	335,241	24.0624	619	383,161	24.8797
580	336,400	24.0832	620	384,400	24.8998
581	337,561	24.1039	621	385,641	24.9199
582	338,724	24.1247	622	386,884	24.9399
583	339,889	24.1454	623	388,129	24.9600
584	341,056	24.1661	624	389,376	24.9800
585	342,225	24.1868	625	390,625	25.0000
586	343,396	24.2074	626	391,876	25.0200
587	344,569	24.2281	627	393,129	25.0400
588	345,744	24.2487	628	394,384	25.0599
589	346,921	24.2693	629	395,641	25.0799
590	348,100	24.2899	630	396,900	25.0998
591	349,281	24.3105	631	398,161	25.1197
592	350,464	24.3311	632	399,424	25.1396
593	351,649	24.3516	633	400,689	25.1595
594	352,836	24.3721	634	401,956	25.1794
595	354,025	24.3926	635	403,225	25.1992
596	355,216	24.4131	636	404,496	25.2190
597	356,409	24.4336	637	405,769	25.2389
598	357,604	24.4540	638	407,044	25.2587
599	358,801	24.4745	639	408,321	25.2784
600	360,000	24.4949	640	409,600	25.2982

TABLE II SQUARES AND SQUARE ROOTS (CONTINUED)

Number	Square	Square root	Number	Square	Square root
641	410,881	25.3180	681	463,761	26.0960
642	412,164	25.3377	682	465,124	26.1151
643	413,449	25.3574	683	466,489	26.1343
644	414,736	25.3772	684	467,856	26.1534
645	416,025	25.3969	685	469,225	26.1725
646	417,316	25.4165	686	470,596	26.1916
647	418,609	25.4362	687	471,969	26.2107
648	419,904	25.4558	688	473,344	26.2298
649	421,201	25.4755	689	474,721	26.2488
650	422,500	25.4951	690	476,100	26.2679
651	423,801	25,5147	691	477,481	26.2869
652	425,104	25.5343	692	478,864	26.3059
653	426,409	25.5539	693	480,249	26.3249
654	427,716	25.5734	694	481,636	26.3439
655	429,025	25.5930	695	483,025	26.3629
656	430,336	25.6125	696	484,416	26.3818
657	431,649	25.6320	697	485,809	26.4008
658	432,964	25.6515	698	487,204	26.4197
659	434,281	25.6710	699	488,601	26.4386
660	435,600	25.6905	700	490,000	26.4575
661	436,921	25.7099	701	491,401	26.4764
662	438,244	25.7294	702	492,804	26.4953
663	439,569	25.7488	703	494,209	26.5141
664	440,896	25.7682	704	495,616	26.5330
665	442,225	25.7876	705	497,025	26.5518
666	443,556	25.8070	706	498,436	26.5707
667	444,889	25.8263	707	499,849	26.5895
668	446,224	25.8457	708	501,264	26.6083
669	447,561	25.8650	709	502,681	26.6271
670	448,900	25.8844	710	504,100	26.6458
671	450,241	25.9037	711	505,521	26.6646
672	451,584	25.9230	712	506,944	26.6833
673	452,929	25.9422	713	508,369	26.7021
674	454,276	25.9615	714	509,796	26.7208
675	455,625	25.9808	715	511,225	26.7395
676	456,976	26.0000	716	512,656	26.7582
677	458,329	26.0192	717	514,089	26.7769
678	459,684	26,0384	718	515,524	26.7955
679	461,041	26.0576	719	516,961	26.8142
680	462,400	26.0768	720	518,400	26.8328

TABLE II SQUARES AND SQUARE ROOTS (CONTINUED)

Number	Square	Square root	Number	Square	Square root
721	519,841	26.8514	761	579,121	27.5862
722	521,284	26.8701	762	580,644	27.6043
723	522,729	26.8887	763	582,169	27.6225
724	524,176	26.9072	764	583,696	27.6405
725	525,625	26.9258	765	585,225	27.6586
726	527,076	26.9444	766	586,756	27.6767
727	528,529	26.9629	767	588,289	27.6948
728	529,984	26.9815	768	589,824	27.7128
729	531,441	27.0000	769	591,361	27.7308
730	532,900	27.0185	770	592,900	27.7489
731	534,361	27.0370	771	594,441	27.7669
732	535,824	27.0555	772	595,984	27.7849
733	537,289	27.0740	773	597,529	27.8029
734	538,756	27.0924	774	599,076	27.8209
735	540,225	27.1109	775	600,625	27.8388
736	541,696	27.1293	776	602,176	27.8568
737	543,169	27.1477	777	603,729	27.8747
738	544,644	27.1662	778	605,284	27.8927
739	546,127	26.1846	779	606,841	27.9106
740	547,600	27.2029	780	608,400	27.9285
741	549,081	27.2213	781	609,961	27.9464
742	550,564	27.2397	782	611,524	27.9643
743	552,049	27.2580	783	613,089	27.9821
744	553,536	27.2764	784	614,656	28.0000
745	555,025	27.2947	785	616,225	28.0179
746	556,516	27.3130	786	617,796	28.0357
747	558,009	27.3313	787	619,369	28.0535
748	559,504	27.3496	788	620,944	28.0713
749	561,001	27.3679	789	622,521	28.0891
750	562,500	27.3861	790	624,100	28.1069
751	564,001	27.4044	791	625,681	28.1247
752	565,504	27.4226	792	627,264	28.1425
753	567,009	27.4408	793	628,849	28.1603
754	568,516	27.4591	794	630,436	28.1780
755	570,025	27.4773	795	632,025	28.1957
756	571,536	27.4955	796	633,616	28.2135
757	573,049	27.5136	797	635,209	28.2312
758	574,564	27.5318	798	636,804	28.2489
759	576,081	27.5500	799	638,401	28.2666
760	577,600	27.5681	800	640,000	28.2843

TABLE II SQUARES AND SQUARE ROOTS (CONTINUED)

Number	Square	Square root	Number	Square	Square root
801	641,601	28.3019	841	707,281	29.0000
802	643,204	28.3196	842	708,964	29.0172
803	644,809	28.3373	843	710,649	29.0345
804	646,416	28.3549	844	712,336	29.0517
805	648,025	28.3725	845	714,025	29.0689
806	649,636	28.3901	846	715,716	29.0861
807	651,249	28.4077	847	717,409	29.1033
808	652,864	28.4253	848	719,104	29.1204
809	654,481	28.4429	849	720,801	29.1376
810	656,100	28.4605	850	722,500	29.1548
811	657,721	28.4781	851	724,201	29.1719
812	659,344	28.4956	852	725,904	29.1890
813	660,969	28.5132	853	727,609	29.2062
814	662,596	28.5307	854	729,316	29.2233
815	664,225	28.5482	855	731,025	29.2404
816	665,856	28.5657	856	732,736	29.2575
817	667,489	28.5832	857	734,449	29.2746
818	669,124	28.6007	858	736,164	29.2916
819	670,761	28.6082	859	737,881	29.3087
820	672,400	28.6356	860	739,600	29.3258
821	674,041	28.6531	861	741,321	29.3428
822	675,684	28.6705	862	743,044	29.3598
823	677,329	28.6880	863	744,769	29.3769
824	678,976	28.7054	864	746,496	29.3939
825	680,625	28.7228	865	748,225	29.4109
826	682,276	28.7402	866	749,956	29.4279
827	683,929	28.7576	867	751,689	29.4449
828	685,584	28.7750	868	753,424	29.4618
829	687,241	28.7924	869	755,161	29.4788
830	688,900	28.8097	870	756,900	29.4958
831	690,561	28.8271	871	758,641	29.5127
832	692,224	28.8444	872	760,384	29.5296
833	693,889	28.8617	873	762,129	29.5466
834	695,556	28.8791	874	763,876	29.5635
835	697,225	28.8964	875	765,625	29.5804
836	698,896	28.9137	876	767,376	29.5973
837	700,569	28.9310	877	769,129	29.6142
838	702,244	28.9482	878	770,884	29.6311
839	703,921	28.9655	879	772,641	29.6479
840	705,600	28.9828	880	774,400	29.6648

TABLE II SQUARES AND SQUARE ROOTS (CONTINUED)

Number	Square	Square root	Number	Square	Square root
881	776,161	29.6816	921	848,241	30.3480
882	777,924	29.6985	922	850,084	30.3645
883	779,689	29.7153	923	851,929	30.3809
884	781,456	29.7321	924	853,776	30.3974
885	783,225	29.7489	925	855,625	30.4138
886	784,996	29.7658	926	857,476	30.4302
887	786,769	29.7825	927	859,329	30.4467
888	788,544	29.7993	928	861,184	30.4631
889	790,321	29.8161	929	863,041	30.4795
890	792,100	29.8329	930	864,900	30.4959
891	793,881	29.8496	931	866,761	30.5123
892	795,664	29.8664	932	868,624	30.5287
893	797,449	29.8831	933	870,489	30.5450
894	799,236	29.8998	934	872,356	30.5614
895	801,025	29.9166	935	874,225	30.5778
896	802,816	29.9333	936	876,096	30.5941
897	804,609	29.9500	937	877,969	30.6105
898	806,404	29.9666	938	879,844	30.6268
899	808,201	29.9833	939	881,721	30.6431
900	810,000	30.0000	940	883,600	30.6594
901	811,801	30.0167	941	885,481	30.6757
902	813,604	30.0333	942	887,364	30.6920
903	815,409	30.0500	943	889,249	30.7083
904	817,216	30.0666	944	891,136	30.7246
905	819,025	30.0832	945	893,025	30.7409
906	820,836	30.0998	946	894,916	30.7571
907	822,649	30.1164	947	896,809	30.7734
908	824,464	30.1330	948	898,704	30.7896
909	826,281	30.1496	949	900,601	30.8085
910	828,100	30.1662	950	902,500	30.8221
911	829,921	30.1828	951	904,401	30.8383
912	831,744	30.1993	952	906,304	30.8545
913	833,569	30.2159	953	908,209	30.8707
914	835,396	30.2324	954	910,116	30.8869
915	837,225	30.2490	955	912,025	30.9031
916	839,056	30.2655	956	913,936	30.9192
917	840.889	30.2820	957	915,849	30.9354
918	842,724	30.2985	958	917,764	30.9516
919	844,561	30.3150	959	919,681	30.9677
920	846,400	30.3315	960	921,600	30.9839

TABLE II SQUARES AND SQUARE ROOTS (CONTINUED)

Number	Square	Square root	Number	Square	Square root
961	923,521	31.0000	981	962,361	31.3209
962	925,444	31.0161	982	964,324	31.3369
963	927,369	31.0322	983	966,289	31.3528
964	929,296	31.0483	984	968,256	31.3688
965	931,225	31.0644	985	970,225	31.3847
966	933,156	31.0805	986	972,196	31.4006
967	935,089	31.0966	987	974,169	31.4166
968	937,024	31.1127	988	976,144	31.4325
969	938,961	31.1288	989	978,121	31.4484
970	940,900	31.1448	990	980,100	31.4643
971	942,841	31.1609	991	982,081	31.4802
972	944,784	31.1769	992	984,064	31.4960
973	946,729	31.1929	993	986,049	31.5119
974	948,676	31.2090	994	988,036	31.5278
975	950,625	31.2250	995	990,025	31.5436
976	952,576	31.2410	996	992,016	31.5595
977	954,529	31.2570	997	994,009	31.5753
978	956,484	31.2730	998	996,004	31.5911
979	958,441	31.2890	999	998,001	31.6070
980	960,400	31.3050	1,000	1,000,000	31.6228

APPENDIX **C**

How to Solve Normal Equations Simultaneously

SIMULTANEOUS EQUATIONS

The word *simultaneous* means "at the same time." Two things happen simultaneously if they happen at the same time. Simultaneous equations have two characteristics:

1. They each have the same two unknowns, usually X and Y.
2. They have one set of values for X and Y which is the same for each.

To solve a pair of simultaneous equations, it is necessary to find the pair of values which satisfies both equations.

One way of accomplishing this is to graph both equations on the same set of axes. The point of intersection of the two straight lines is the point of simultaneous equality. The coordinates of this point are the X and Y values which satisfy both equations.

In addition to the graphic method, simultaneous equations may be solved algebraically. In statistics, the algebraic solution is used in order to obtain more accurate results.

The procedure is as follows:

1. If the coefficients of one unknown in both equations are not the same, multiply both members of the equations by numbers that make the coefficients of one unknown numerically equal.
2. Eliminate one unknown from the equations. This is done by adding or subtracting the new equations to obtain a single equation in one unknown.

3. Solve this equation for the unknown.
4. Find the other unknown by either:
   a. Substituting the value obtained for the first unknown in either of the original equations, or
   b. Repeating steps 1, 2, 3, and eliminating the other unknown.
5. Check answer by substituting the values in both original equations.

EXAMPLE

Given the data in Table C.1, compute the regression line by the method of least squares:

TABLE C.1

X	Y
1	7
2	11
3	14
4	13
5	18
6	24
7	25
8	28

Solution: Two normal equations are solved simultaneously to derive the line of regression. The following sums must be computed: ΣX, ΣY, ΣXY, ΣX^2. Therefore, Table C.2 is set up.

TABLE C.2

X	Y	XY	X^2
1	7	7	1
2	11	22	4
3	14	42	9
4	13	52	16
5	18	90	25
6	24	144	36
7	25	175	49
8	28	224	64
36	140	756	204

The two normal equations are:

$$\begin{aligned} &\text{(I)} \quad \Sigma Y = na + b\Sigma X \\ &\text{(II)} \quad \Sigma XY = a\Sigma X + b\Sigma X^2 \end{aligned}$$

1. From the table above, substitute values in the normal equations as follows:

$$\begin{aligned} &\text{(I)} \quad 140 = 8a + 36b \\ &\text{(II)} \quad 756 = 36a + 204b \end{aligned}$$

2. To eliminate a and solve for b, multiply each item in equation 1 by -4.5. The value 4.5 is derived by dividing 36 by 8, which equals 4.5. The minus sign is assigned to 4.5 in order to permit algebraic eliminations of a.

$$\begin{aligned} &\text{(I)} \quad 140 = 8a + 36b \quad (-4.5) \\ &\text{(II)} \quad 756 = 36a + 204b \end{aligned}$$

$$\begin{aligned} &\text{(I)} \quad -630 = -36a - 162b \\ &\text{(II)} \quad 756 = 36a + 204b \end{aligned}$$

$$\begin{aligned} 126 &= 42b \\ 42b &= 126 \\ b &= 3 \end{aligned}$$

3. Substitute b value in equation I:

$$\begin{aligned} \text{(I)} \quad 140 &= 8a + 36b \\ 140 &= 8a + (36)(3) \\ 140 &= 8a + 108 \\ -8a &= -140 + 108 \\ +8a &= +32 \\ a &= 4 \end{aligned}$$

4. Substitute computed values for a and b in the line-of-regression formula, $Y_c = a + bX$: Thus, $Y_c = 4 + 3X$.

5. To check the accuracy of the computations, substitute a and b values in each of the two normal equations:

$$\begin{aligned} \text{(I)} \quad 140 &= (8)(4) + (36)(3) \\ 140 &= 32 + 108 \\ 140 &= 140 \end{aligned}$$

$$\begin{aligned} \text{(II)} \quad 756 &= (36)(4) + (204)(3) \\ 756 &= 144 + 612 \\ 756 &= 756 \end{aligned}$$

6. Compute Y_c values for the formula for the line of regression. For each X value, it is now possible to obtain a computed Y or Y_c, value (Table C.3).

TABLE C.3

X	Y_c
1	7
2	10
3	13
4	16
5	19
6	22
7	25
8	28

If $X = 1$, then $Y_c = 4 + (3)(1)$
$= 4 + 3$
$= 7$

If $X = 2$, then $Y_c = 4 + (3)(2)$
$= 4 + 6$
$= 10$

If $X = 3$, then $Y_c = 4 + (3)(3)$
$= 4 + 9$
$= 13$, and so forth. Thus

If $X = 4$, $Y_c = 16$
If $X = 5$, $Y_c = 19$
If $X = 6$, $Y_c = 22$
If $X = 7$, $Y_c = 25$
If $X = 8$, $Y_c = 28$

7. To draw a straight line requires only two points. If it is necessary to plot the line of regression on a graph, it is enough to locate, say, coordinates (1, 7) and coordinates (8, 28) on the graph and to draw a straight line through these points.

APPENDIX **D**

Selected Sources of Economic and Business Data

GENERAL GUIDES

Chamber of Commerce of the U.S., *What's the Answer?*, Washington, D.C., 1953.

Coman, Jr., Edwin T., *Sources of Business Information*, Englewood Cliffs, N.J., Prentice-Hall, 1949.

Johnson, H. Webster, and Stuart W. McFarland, *How to Use the Business Library, with Sources of Business Information*, 2nd ed., Cincinnati, South-Western, 1957.

Manley, Marian C., *Business Information: How to Find and Use it*, New York, Harper and Row, 1955.

Wasserman, Paul, *Information for Administrators*, Ithaca, N.Y., Cornell University Press, 1956.

Wasserman, Paul *et al.* (ed.), *Statistics Sources*, Detroit, Gale, 1962.

SPECIALIZED GUIDES

Hauser, Philip M., and William R. Leonard (ed.), *Government Statistics for Business Use*, New York, John Wiley, 1956.

Cole, Arthur, H., *Measures of Business Change*, Chicago, Irwin, 1952.

INDEXES

Business Periodicals Index, New York, H. W. Wilson, 1958 to date.

Monthly Catalog of United States Government Publications, Washington, D.C., Government Printing Office, 1895 to date.

New York Times Index, New York, New York Times, 1913 to date.

Public Affairs Information Service, New York, Public Affairs Information Service, 1915 to date.

Wall Street Journal Index, New York, Wall Street Journal, 1958 to date.

SELECTED DATA SOURCES

Moody's Investors Service, *Moody's Manuals*, New York, 5 annual volumes: *Industrials*, *Public Utilities*, *Transportation*, *Governments*, and *Banks-Insurance-Real Estate*.

Standard and Poor's Corporation, *Standard Corporation Records*, New York. Current data on over 6,000 corporations.

U.S. Department of Agriculture, *Agricultural Statistics*, U.S. Government Printing Office, 1936, to date. Annual publication.

U.S. Department of Commerce and Board of Governors of the Federal Reserve System, *Banking and Monetary Statistics*, 1913–1941, U.S. Government Printing Office, 1943.

U.S. Department of Commerce and Board of Governors of the Federal Reserve System, *Federal Reserve Bulletin*. Monthly publication, U.S. Government Printing Office, current data.

U.S. Department of Commerce, Bureau of the Census, *Statistical Abstract of the United States*, U.S. Government Printing Office, 1878 to date. Annual publication.

———, *Historical Statistics of the United States: Colonial Times to* 1957. U.S. Government Printing Office, 1960.

———, *Census of Business* and *Census of Manufactures*. Quinquennial publications.

U.S. Department of Labor, Bureau of Labor Statistics, *Handbook of Labor Statistics*. 1926–1951, U.S. Government Printing Office.

———, *Monthly Labor Review*. Monthly publication. U.S. Government Printing Office, current data.

TABLE III AREAS UNDER THE NORMAL CURVE

(Proportion of total area within stated interval away from the mean on one side of the mean)

z	.00	.01	.02	.03	.04	.05	.06	.07	.08	.09
0.0	00000	00399	00798	01197	01595	01994	02392	02790	03188	03586
0.1	03983	04380	04776	05172	05567	05962	06356	06749	07142	07535
0.2	07926	08317	08706	09095	09483	09871	10257	10642	11026	11409
0.3	11791	12172	12552	12930	13307	13683	14058	14431	14803	15173
0.4	15542	15910	16276	16640	17003	17364	17724	18082	18439	18793
0.5	19146	19497	19847	20194	20540	20884	21226	21566	21904	22240
0.6	22575	22907	23237	23565	23891	24215	24537	24857	25175	25490
0.7	25804	26115	26424	26730	27035	27337	27637	27935	28230	28524
0.8	28814	29103	29389	29673	29955	30234	30511	30785	31057	31327
0.9	31594	31859	32121	32381	32639	32894	33147	33398	33646	33891
1.0	34134	34375	34614	34850	35083	35314	35543	35769	35993	36214
1.1	36433	36650	36864	37076	37286	37493	37698	37900	38100	38298
1.2	38493	38686	38877	39065	39251	39435	39617	39796	39973	40147
1.3	40320	40490	40658	40824	40988	41149	41309	41466	41621	41774
1.4	41924	42073	42220	42364	42507	42647	42786	42922	43056	43189
1.5	43319	43448	43574	43699	43822	43943	44062	44179	44295	44408
1.6	44520	44630	44738	44845	44950	45053	45154	45254	45352	45449
1.7	45543	45637	45728	45818	45907	45994	46080	46164	46246	46327
1.8	46407	46485	46562	46638	46712	46784	46856	46926	46995	47062
1.9	47128	47193	47257	47320	47381	47441	47500	47558	47615	47670
2.0	47725	47778	47831	47882	47932	47982	48030	48077	48124	48169
2.1	48214	48257	48300	48341	48382	48422	48461	48500	48537	48574
2.2	48610	48645	48679	48713	48745	48778	48809	48840	48870	48899
2.3	48928	48956	48983	49010	49036	49061	49086	49111	49134	49158
2.4	49180	49202	49224	49245	49266	49286	49305	49324	49343	49361
2.5	49379	49396	49413	49430	49446	49461	49477	49492	49506	49520
2.6	49534	49547	49560	49573	49585	49598	49609	49621	49632	49643
2.7	49653	49664	49674	49683	49693	49702	49711	49720	49728	49736
2.8	49744	49752	49760	49767	49774	49781	49788	49795	49801	49807
2.9	49813	49819	49825	49831	49836	49841	49846	49851	49856	49861
3.0	49865	49869	49874	49878	49882	49886	49889	49893	49897	49900
3.1	49903	49906	49910	49913	49916	49918	49921	49924	49926	49929

TABLE IV RANDOM DIGITS

	1	2	3	4	5	6	7	8	9	10
1	10480	15011	01536	02011	81647	91646	69179	14194	62590	36207
2	22368	46573	25595	85393	30995	89198	27982	53402	93965	34095
3	24130	48360	22527	97265	76393	64809	15179	24830	49340	32081
4	42167	93093	06243	61680	07856	16376	39440	53537	71341	57004
5	37570	39975	81837	16656	06121	91782	60468	81305	49684	60672
6	77921	06907	11008	42751	27756	53498	18602	70659	90655	15053
7	99562	72905	56420	69994	98872	31016	71194	18738	44013	48840
8	96301	91977	05463	07972	18876	20922	94595	56869	69014	60045
9	89579	14342	63661	10281	17453	18103	57740	84378	25331	12566
10	85475	36857	53342	53988	53060	59533	38867	62300	08158	17983
11	28918	69578	88231	33276	70997	79936	56865	05859	90106	31595
12	65553	40961	48235	03427	49626	69445	18663	72695	52180	20847
13	09429	93969	52636	92737	88974	33488	36320	17617	30015	08272
14	10365	61129	87529	85689	48237	52267	67689	93394	01511	26358
15	07119	97336	71048	08178	77233	13916	47564	81056	97735	85977
16	51085	12765	51821	51259	77452	16308	60756	92144	49442	53900
17	02368	21382	52404	60268	89368	19885	55322	44819	01188	65255
18	01011	54092	33362	94904	31273	04146	18594	29852	71585	85030
19	52162	53916	46369	58586	23216	14513	83149	98736	23495	64350
20	07056	97628	33787	09998	42698	06691	76988	13602	51851	46104
21	48663	91245	85828	14346	09172	30168	90229	04734	59193	22178
22	54164	58492	22421	74103	47070	25306	76468	26384	58151	06646
23	32639	32363	05597	24200	13363	38005	94342	28728	35806	06912
24	29334	27001	87637	87308	58731	00256	45834	15398	46557	41135
25	02488	33062	28834	07351	19731	92420	60952	61280	50001	67658
26	81525	72295	04839	96423	24878	82651	66566	14778	76797	14780
27	29676	20591	68086	26432	46901	20849	89768	81536	86645	12659
28	00742	57392	39064	66432	84673	40027	32832	61362	98947	96067
29	05366	04213	25669	26422	44407	44048	37937	63904	45766	66134
30	91921	26418	64117	94305	26766	25940	39972	22209	71500	64568
31	00582	04711	87917	77341	42206	35126	74087	99547	81817	42607
32	00725	69884	62797	56170	86324	88072	76222	36086	84637	93161
33	69011	65795	95876	55293	18988	27354	26575	08625	40801	59920
34	25976	57948	29888	88604	67917	48708	18912	82271	65424	69774
35	09763	83473	73577	12908	30883	18317	28290	35797	05998	41688
36	91567	42595	27958	30134	04024	86385	29880	99730	55536	84855
37	17955	56349	90999	49127	20044	59931	06115	20542	18059	02008
38	46503	18584	18845	49618	02304	51038	20655	58727	28168	15475
39	92157	89634	94824	78171	84610	82834	09922	25417	44137	48413
40	14577	62765	35605	81263	39667	47358	56873	56307	61607	49518
41	98427	07523	33362	64270	01638	92477	66969	98420	04880	45585
42	34914	63976	88720	82765	34476	17032	87589	40836	32427	70002
43	70060	28277	39475	46473	23219	53416	94970	25832	69975	94884
44	53976	54914	06990	67245	68350	82948	11398	42878	80287	88267
45	76072	29515	40980	07391	58745	25774	22987	80059	39911	96189
46	90725	52210	83974	29992	65831	38857	50490	83765	55657	14361
47	64364	67412	33339	31926	14883	24413	59744	92351	97473	89236
48	08962	00358	31662	25388	61642	34072	81249	35648	56891	69352
49	95012	68379	93526	70765	10592	04542	76463	54328	02349	17247
50	15664	10493	20492	38391	91132	21999	59516	81652	27195	48223

SOURCE: Interstate Commerce Commission, *Table of 105,000 Random Decimal Digits*, Washington, D.C., May 1949, p. 1.

TABLE V DISTRIBUTION OF *t*

Degrees of freedom (n)	Probability		
	.10	.05	.01
1	6.314	12.706	63.657
2	2.920	4.303	9.925
3	2.353	3.182	5.841
4	2.132	2.776	4.604
5	2.015	2.571	4.032
6	1.943	2.447	3.707
7	1.895	2.365	3.499
8	1.860	2.306	3.355
9	1.833	2.262	3.250
10	1.812	2.228	3.169
11	1.796	2.201	3.106
12	1.782	2.179	3.055
13	1.771	2.160	3.012
14	1.761	2.145	2.977
15	1.753	2.131	2.947
16	1.746	2.120	2.921
17	1.740	2.110	2.898
18	1.734	2.101	2.878
19	1.729	2.093	2.861
20	1.725	2.086	2.845
21	1.721	2.080	2.831
22	1.717	2.074	2.819
23	1.714	2.069	2.807
24	1.711	2.064	2.797
25	1.708	2.060	2.787
26	1.706	2.056	2.779
27	1.703	2.052	2.771
28	1.701	2.048	2.763
29	1.699	2.045	2.756
30	1.697	2.042	2.750
∞	1.645	1.960	2.576

SOURCE: Ronald Fisher and F. Yates, *Statistical Tables for Biological, Agricultural and Medical Research*, 4th ed., Oliver and Boyd, Edinburgh, Eng., 1953, Table IV. Reprinted by permission of the authors and publishers.

TABLE VI CHI-SQUARE VALUES

Degrees of freedom (n)	Probability		
	0.10	0.05	0.01
1	2.706	3.841	6.635
2	4.605	5.991	9.201
3	6.251	7.815	11.341
4	7.779	9.488	13.277
5	9.236	11.070	15.086
6	10.645	12.592	16.812
7	12.017	14.067	18.475
8	13.362	15.507	20.090
9	14.684	16.919	21.666
10	15.987	18.307	23.209
11	17.275	19.675	24.725
12	18.549	21.026	26.217
13	19.812	22.362	27.688
14	21.064	23.685	29.141
15	22.307	24.996	30.578
16	23.542	26.296	32.000
17	24.769	27.587	33.409
18	25.989	28.869	34.805
19	27.204	30.144	36.191
20	28.412	31.410	37.566
21	29.615	32.671	38.932
22	30.813	33.924	40.289
23	32.007	35.172	41.638
24	33.196	36.415	42.980
25	34.382	37.652	44.314
26	35.563	38.885	45.642
27	36.741	40.113	46.963
28	37.916	41.337	48.278
29	39.088	42.557	49.588
30	40.256	43.773	50.892

SOURCE: Ronald Fisher and F. Yates, *Statistical Tables for Biological, Agricultural and Medical Research*, 4th ed., Oliver and Boyd, Edinburgh, Eng., 1953, Table III. Reprinted by permission of the authors and publishers.

Bibliography

PRESENTATION OF DATA

Jenkinson, Bruce L., *Bureau of the Census Manual of Tabular Presentation*, Washington, D.C., U.S. Government Printing Office, 1949.
Lutz, R. R., *Graphic Presentation Simplified*, New York, Funk and Wagnalls, 1949.
Modley, Rudolph, *How to Use Pictorial Statistics*, New York, Harper, 1937.
Modley, Rudolph, and D. Lowenstein, *Pictographs and Graphs*, New York, Harper and Row, 1952.
Myers, John H., *Statistical Presentation*, Paterson, N.J., Littlefield, Adams, 1956.
Schmid, Calvin F., *Handbook of Graphic Presentation*, New York, Ronald, 1954.
Smart, L. E., and S. Arnold, *Practical Rules for Graphic Presentation of Business Statistics*, Columbus, Ohio, Ohio State University, 1951.
Spear, M. E., *Charting Statistics*, New York, McGraw-Hill, 1952.

PROBABILITY AND STATISTICAL INFERENCE

Bross, I. D. J., *Design for Decision*, New York, Macmillan, 1953.
Burington, Richard S., and D. C. May, *Handbook of Probability and Statistics*, New York, McGraw-Hill, 1953.
Cochran, W. G., *Sampling Techniques*, New York, John Wiley, 1953.
Deming, William E., *Some Theory of Sampling*, New York, John Wiley, 1950.
———, *Sample Design in Business Research*, New York, John Wiley, 1960.
Feigenbaum, A. V., *Quality Control: Principles, Practice, and Administration*, New York, McGraw-Hill, 1951.

Fisher, Ronald A., *Statistical Methods for Research Workers*, New York, Hafner, 1948.

———, *The Design of Experiments*, New York, Hafner, 1951.

Freund, John E., Paul E. Livermore, and Irwin Miller, *Manual of Experimental Statistics*, Englewood Cliffs, N.J., Prentice-Hall, 1960.

Goldberg, S., *Probability: An Introduction*, Englewood Cliffs, N.J., Prentice-Hall, 1960.

Grant, Eugene L., *Statistical Quality Control*, 2nd ed., New York, McGraw-Hill, 1952.

Hansen, Morris H., William N. Hurwitz, and William G. Madow, *Sample Survey Methods and Theory*, New York, John Wiley, 1953.

Hodges, J. L., and E. L. Lehmann, *Basic Concepts of Probability and Statistics*, San Francisco, Holden-Day, 1964.

Levinson, Horace E., *Chance, Luck and Statistics* (formerly titled *The Science of Chance*), New York, Dover, 1963.

Lindgren, B. W., and G. W. McElrath, *Introduction to Probability and Statistics* New York, Macmillan, 1959.

McCarthy, P. J., *Introduction to Statistical Reasoning*, New York, McGraw-Hill, 1957.

Mosteller, Frederick, Robert E. K. Rourke, and George B. Thomas, Jr., *Probability and Statistics*, Reading, Mass., Addison-Wesley, 1961.

Nagel, E., *Principles of the Theory of Probability*, Chicago, University of Chicago Press, 1939.

Parten, Mildred B., *Surveys, Polls, and Samples: Practical Procedures*, New York, Harper and Row, 1950.

Schlaifer, Robert, *Probability and Statistics for Business Decisions*, New York, McGraw-Hill, 1959.

Vance, Lawrence L., and John Neter, *Statistical Sampling for Auditors and Accountants*, New York, John Wiley, 1956.

Walker, Helen M., and Joseph Lev, *Statistical Inference*, New York, Holt, Rinehart and Winston, 1951.

Weiss, Lionel, *Statistical Decision Theory*, New York, McGraw-Hill, 1961.

REGRESSION AND LINEAR CORRELATION

Ezekiel, Mordecai, and Karl A. Fox, *Methods of Correlation and Regression Analysis*, New York, John Wiley, 1959.

Kendall, M. G., *Rank Correlation Methods*, London, Charles Griffin, 1948.

Schultz, Henry, *The Theory and Measurement of Demand*, Chicago, University of Chicago Press, 1938.

TIME SERIES ANALYSIS AND FORECASTING

Abramson, Adolph G., and Russell H. Mack (ed.), *Business Forecasting in Practice: Principles and Cases*, New York, John Wiley, 1956.

Burns, Arthur F., *Production Trends in the United States Since* 1870, New York, National Bureau of Economic Research, 1934.

Burns, Arthur F., and Wesley C. Mitchell, *Measuring Business Cycles*, New York, National Bureau of Economic Research, 1946.

Controllership Foundation, Inc., *Business Forecasting: A Survey of Business Practices and Methods*, New York, 1950.

Davis, Harold, T., *The Analysis of Economic Time Series*, Bloomington, Ind., Principia, 1941.

Hannan, E. J., *Time Series Analysis*, New York, John Wiley, 1960.

Kuznets, Simon, *Seasonal Variations in Industry and Trade*, New York, National Bureau of Economic Research, 1933.

Macaulay, Frederick R., *The Smoothing of Time Series*, New York, National Bureau of Economic Research, 1931.

Moore, Geoffrey H., *Business Cycle Indicators*, Princeton, N.J., Princeton University Press, 1961.

National Bureau of Economic Research, Conference on Research in Income and Wealth, *Short-Term Economic Forecasting*, Princeton, N.J., Princeton University Press, 1955.

———, *Long-Term Economic Projections*, Princeton, N.J., Princeton University Press, 1954.

Newbury, Frank D., *Business Forecasting: Principles and Practices*, New York, McGraw-Hill, 1952.

Prochnow, Herbert, V. (ed.), *Determining the Business Outlook*, New York, Harper and Row, 1954.

Shiskin, Julius, *Electronic Computers and Business Indicators*, New York, National Bureau of Economic Research, 1957.

Snyder, Richard M., *Measuring Business Changes*, New York, John Wiley, 1955.

Spencer, Milton H., Colin G. Clark, and Peter W. Hoguet, *Business and Economic Forecasting*, Homewood, Ill., Irwin, 1961.

Steiner, Peter O., *An Introduction to the Analysis of Time Series*, New York, Holt, Rinehart and Winston, 1956.

Wright, Wilson, *Forecasting for Profit: A Technique for Business Management*, New York, John Wiley, 1947.

INDEX NUMBERS

Fisher, Irvine, *The Making of Index Numbers*, Boston, Houghton Mifflin, 1923.

Government Price Statistics: Hearings Before the Subcommittee on Economic Statistics, Joint Economic Committee, 87th Cong., 1st Sess., Washington, D.C., U.S. Government Printing Office, 1961.

Hauser, Phillip M., and William R. Leonard, *Government Statistics for Business Use*, 2nd ed., New York, John Wiley, 1956.

Kaplan, Lawrence J., "A Guide to the Federal Government's Indexes of Wholesale Prices," *Analysts Journal*, February 1957.

Mitchell, Wesley C., *The Making and Using of Index Numbers*, Washington, D.C., U.S. Government Printing Office, 1938.

Mudgett, Bruce D., *Index Numbers*, New York, John Wiley, 1951.

Persons, Warren M., *The Construction of Index Numbers*, Boston, Houghton Mifflin, 1928.

United Nations, *Index Numbers of Industrial Production*, New York, 1950.

MISCELLANEOUS

Arkin, Herbert, and Raymond R. Colton, *Tables for Statisticians*, New York, Barnes and Noble, 1950.

Hollingdale, S. H., *High Speed Computing Methods and Applications*, New York, Macmillan, 1959.

Huff, Darrell, *How to Lie With Statistics*, New York, W. W. Norton, 1954.

Interstate Commerce Commission, *Table of* 105,000 *Random Decimal Digits*, Washington, D.C., U.S. Government Printing Office, 1949.

Kendall, Maurice G., and B. B. Smith, *Tables of Random Numbers*, Tracts for Computers No. XXIV, Cambridge, England, Cambridge University Press, 1939.

Kendall, Maurice G., and William R. Buckland, *A Dictionary of Statistical Terms*, New York, Hafner, 1957.

McCracken, Daniel D., Harold Weiss, and Tsai-Hwa Lee, *Programming Business Computers*, New York, John Wiley, 1959.

Rand Corporation, *A Million Random Digits with* 100,000 *Normal Deviates*, New York, Free Press of Glencoe, 1955.

Tintner, G., *Econometrics*, New York, John Wiley, 1952.

Tippett, L. H. C., *Random Sampling Numbers*, Tracts for Computers No. XV, Cambridge, England, Cambridge University Press, 1927.

Walker, Helen M., *Mathematics Essential for Elementary Statistics*, rev. ed., New York, Holt, Rinehart and Winston, 1951.

Index